Level K

Mathematics

2nd Edition

Nancy L. McGraw

Bright Ideas Press, LLC
Cleveland, Ohio

Simple Solutions Level K Second Edition

Printed in the United States of America

ISBN-13: 978-1-934210-29-1
ISBN-10: 1-934210-29-3

Cover Design: Dan Mazzola
Editor: Kimberly A. Dambrogio

Welcome to Simple Solutions

Note to the Student:

Hello,

This year you will learn many new and exciting things in math. This book will help you practice those new things every day so you don't forget them.

You must do the pages in this book every night if you want to be a good math student. If you don't know how to do a problem, ask your teacher for help.

We know you are going to like using this book. Wait until you see how fun math can be!

Lesson #1

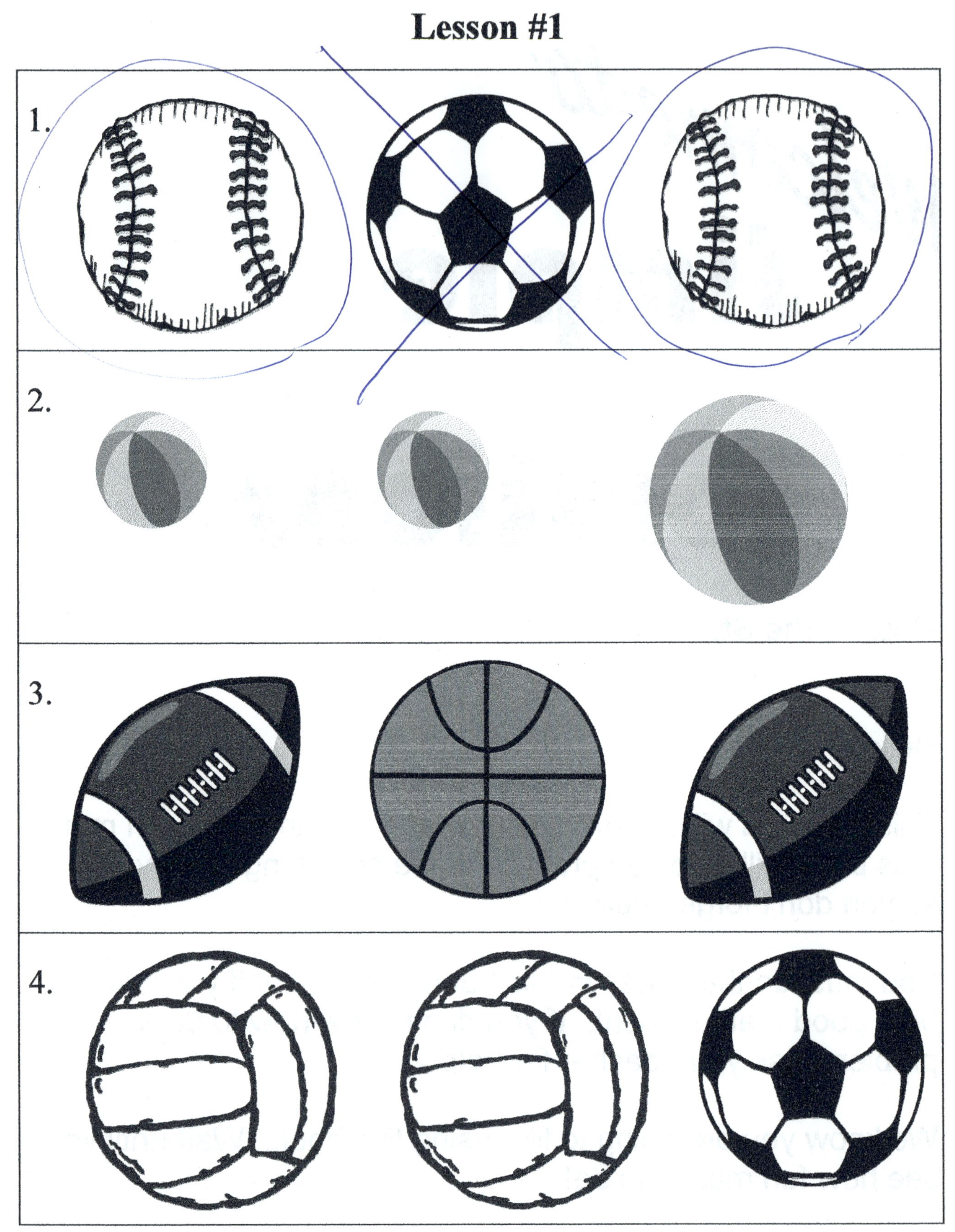

Directions:

1 - 4) Circle the ones that are alike. Cross out the one that is different.

Lesson #2

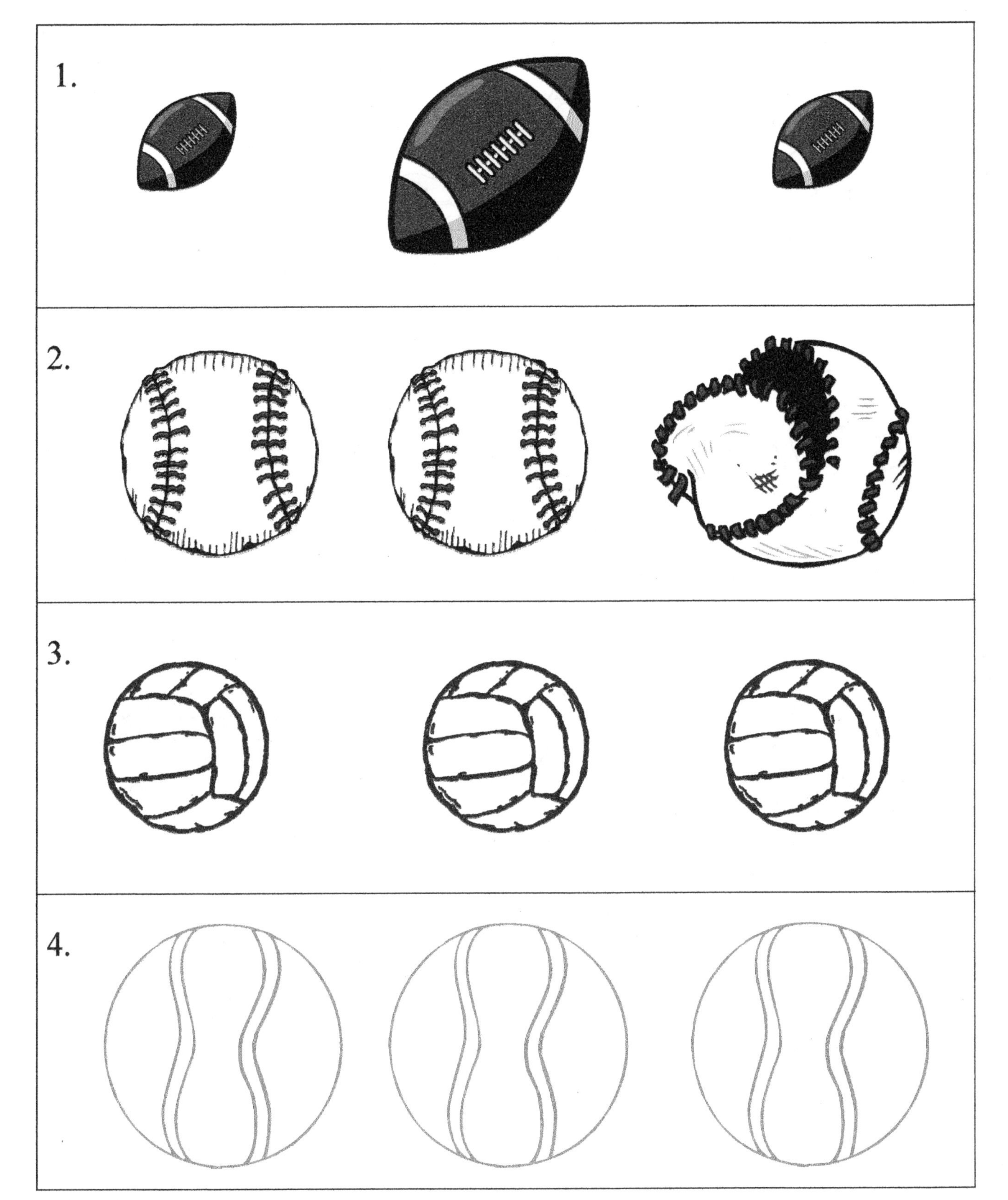

Directions:

1 – 2) Circle the ones that are alike. Cross out the one that is different.

3 – 4) Color 2 balls green and one ball yellow. Then, circle the balls that are alike.

Lesson #3

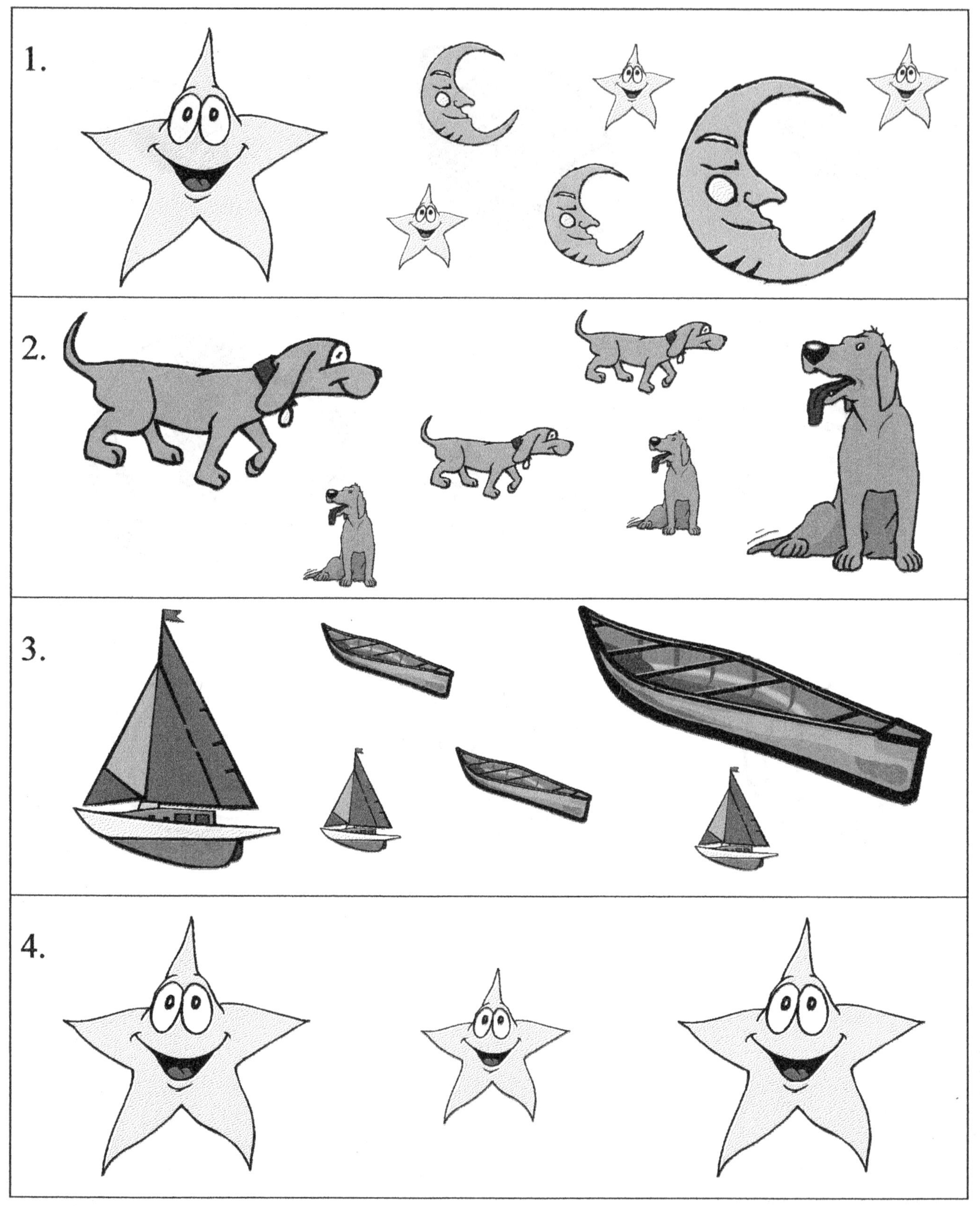

Directions:

1 – 3) Circle the big shapes with a blue crayon. Circle the small shapes with a red crayon.
4) Cross out the one that is different.

Lesson #4

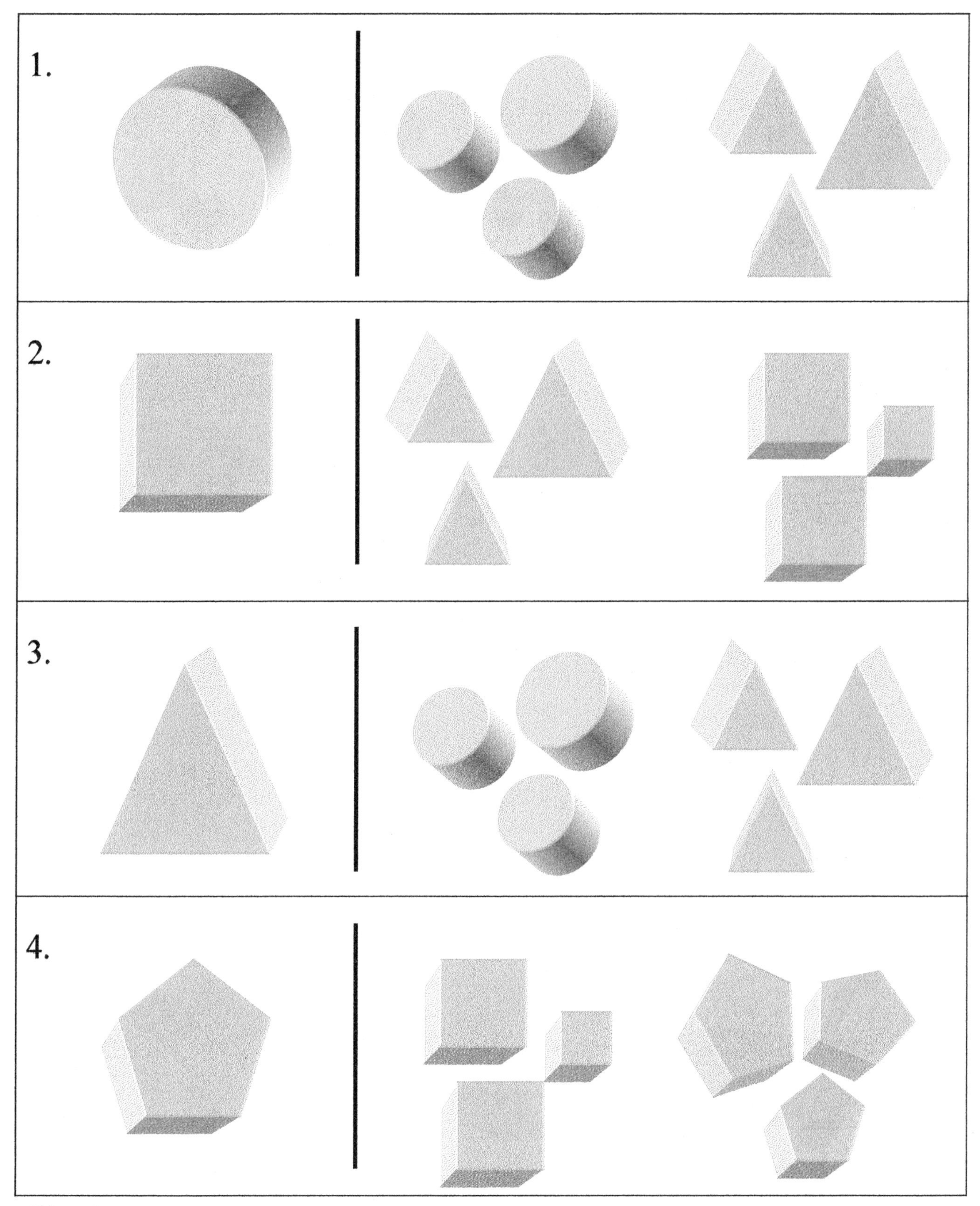

Directions:

1 – 4) Look at the shape before the line. Then, circle the group where the shape belongs.

Lesson # 5

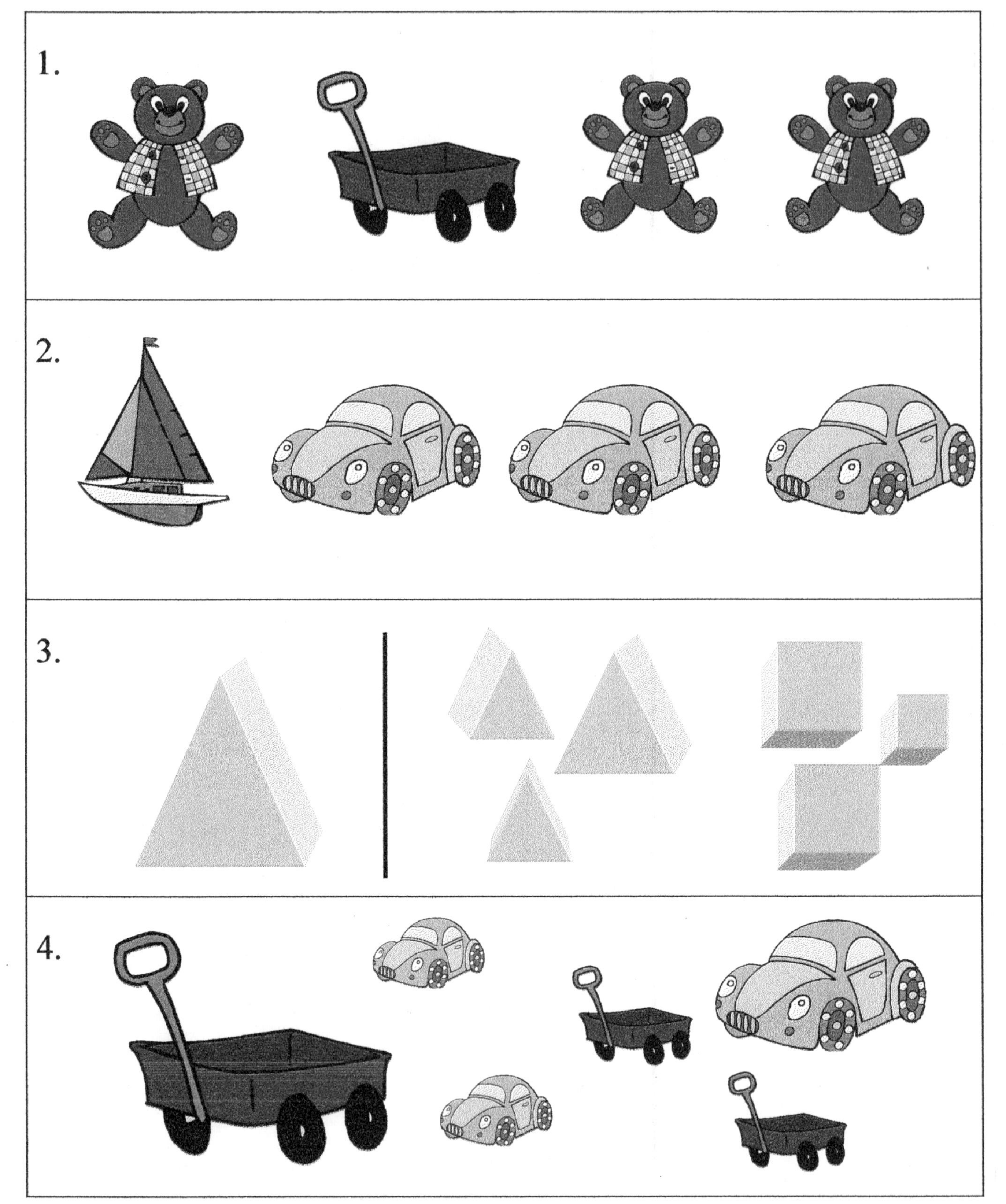

Directions:

1 - 2) Cross out the one that does not belong.

3) Look at the shape before the line. Then, circle the group where the shape belongs.

4) Circle the small shapes with a green crayon.

Lesson #6

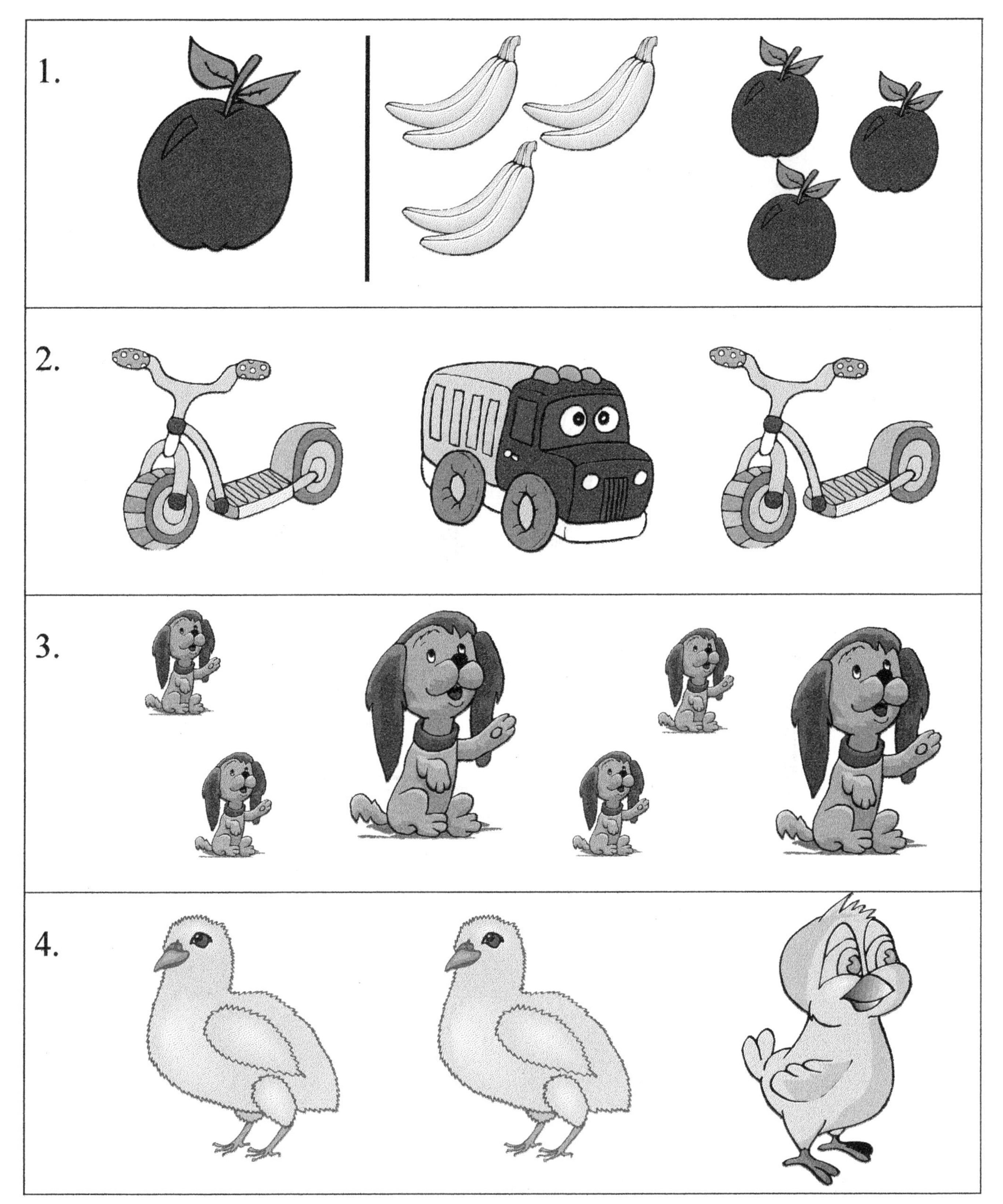

Directions:

1) Look at the object before the line. Then, circle the group where the object belongs.
2) Cross out the one that does not belong.
3) Circle the big dogs in red and the little dogs in blue.
4) Cross out the one that doesn't belong.

Lesson #7

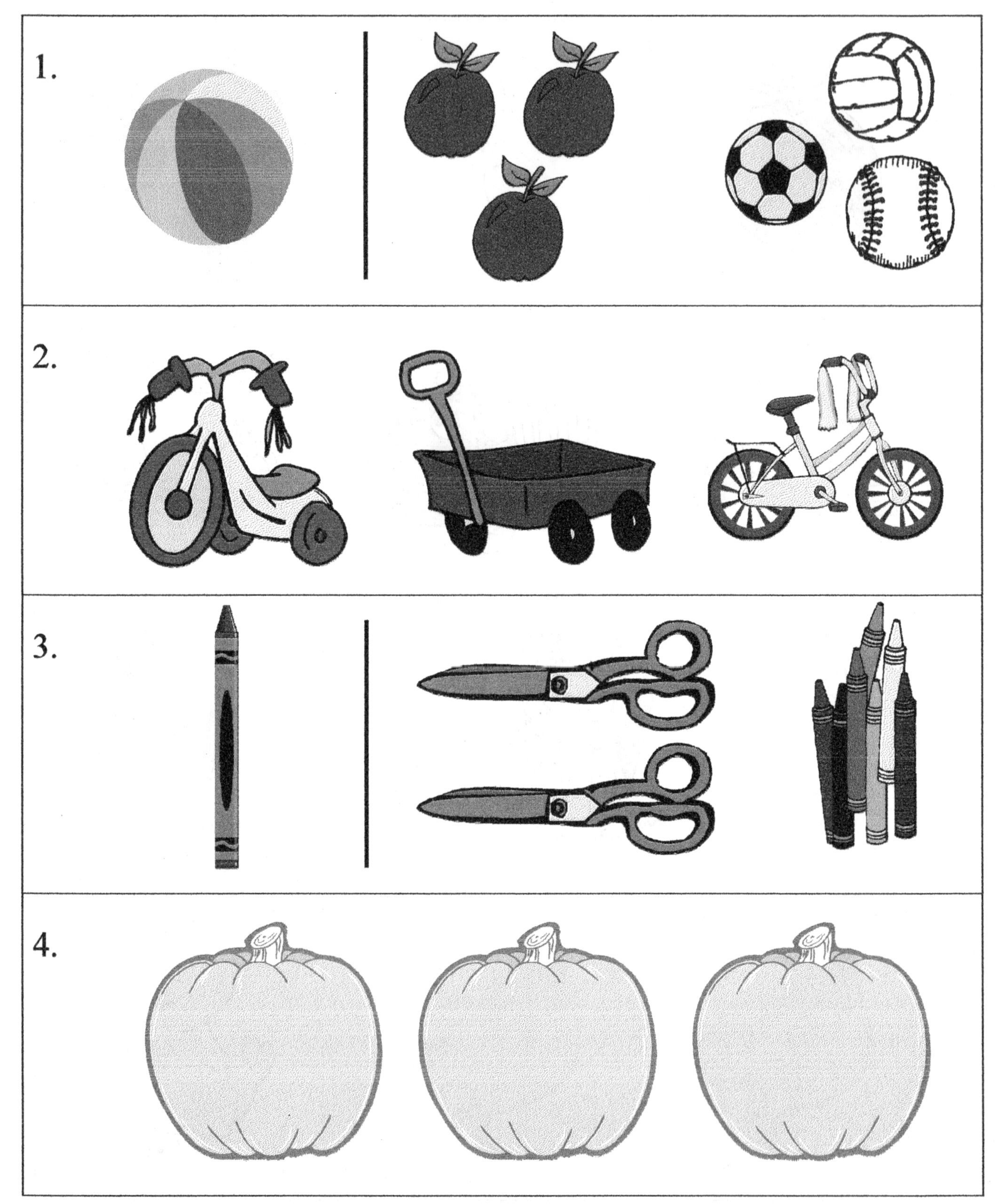

Directions:

1) Look at the object before the line. Then, circle the group where the object belongs.
2) Cross out the one that doesn't belong.
3) Look at the object before the line. Then, circle the group where the object belongs.
4) Color two pumpkins orange. Then, cross out the one that doesn't belong.

Lesson #8

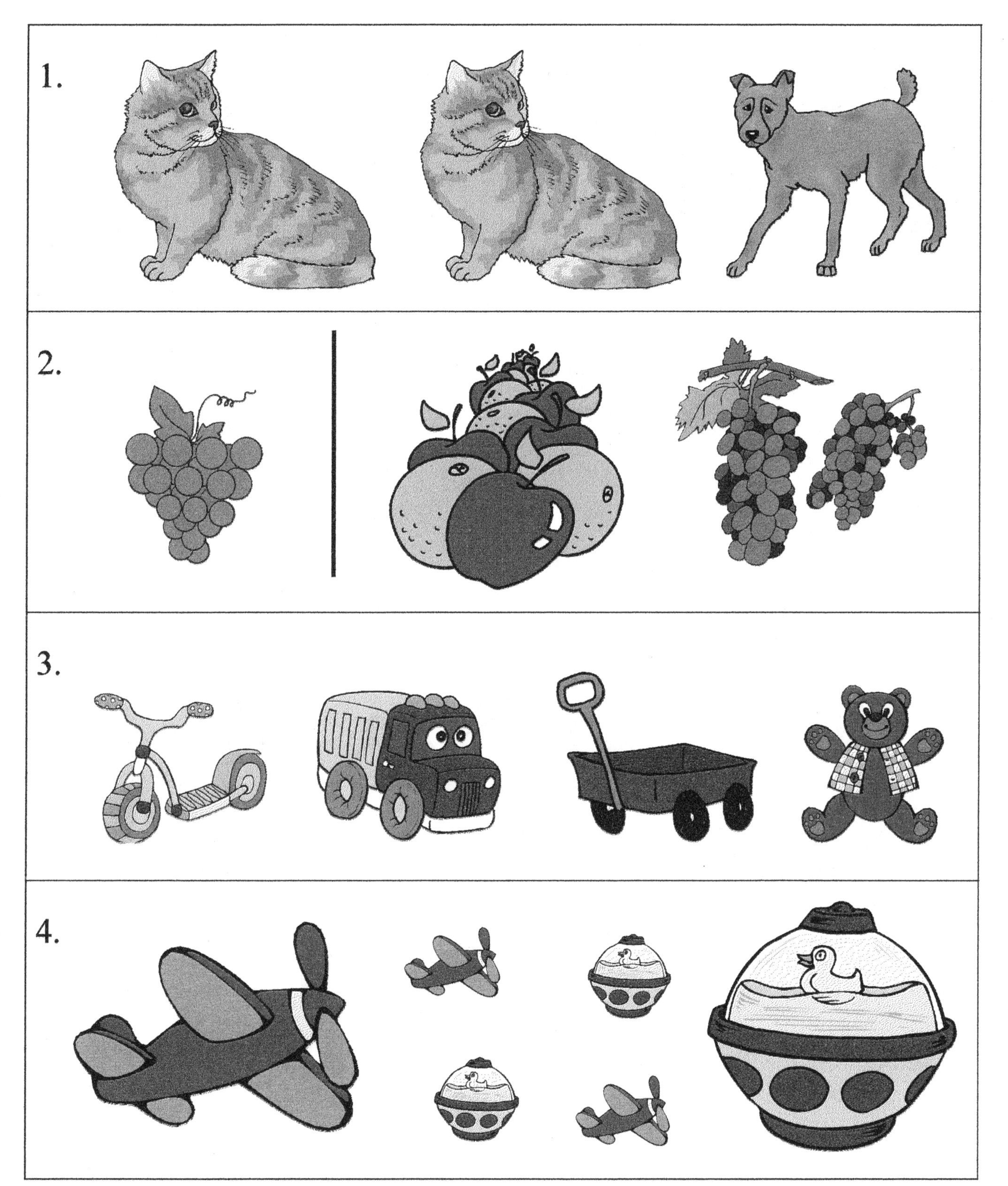

Directions:

1) Cross out the one that doesn't belong.
2) Circle the group where the object belongs.
3) Circle the one that doesn't belong and tell why.
4) Circle the big objects green.

Lesson #9

Directions:

1) Circle the group where the object belongs and tell why.
2) Cross out the one that doesn't belong and tell why.
3) Circle the two fish that are alike.
4) Cross out the one that doesn't belong and tell why.

Lesson #10

Directions:

1 – 2) Look at the object before the line. Then circle the group where it belongs.
3) Cross out the one that doesn't belong with the others.
4) Circle the small objects using an orange crayon.

Lesson #11

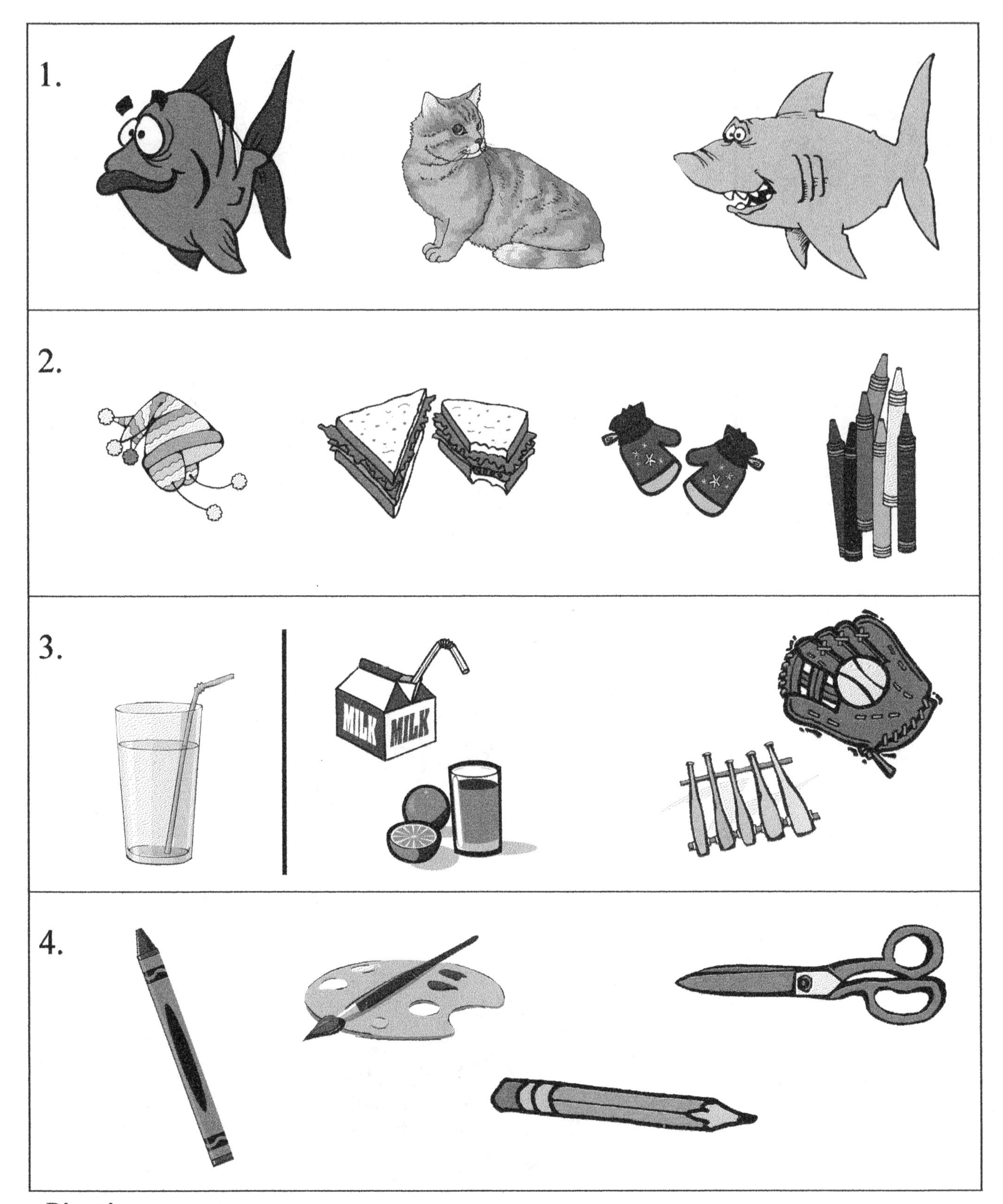

Directions:

1) Cross out the one that doesn't belong.
2) Circle the things that a person can wear.
3) Circle the group where the object belongs.
4) Cross out the things that a person can use to draw.

Lesson #12

1.
2.
3.
4.

Directions:

1) Cross out the one that doesn't belong.
2) Circle the group where the object belongs.
3) Circle the things a person can wear on their feet.
4) Circle the big mice in blue.

Lesson #13

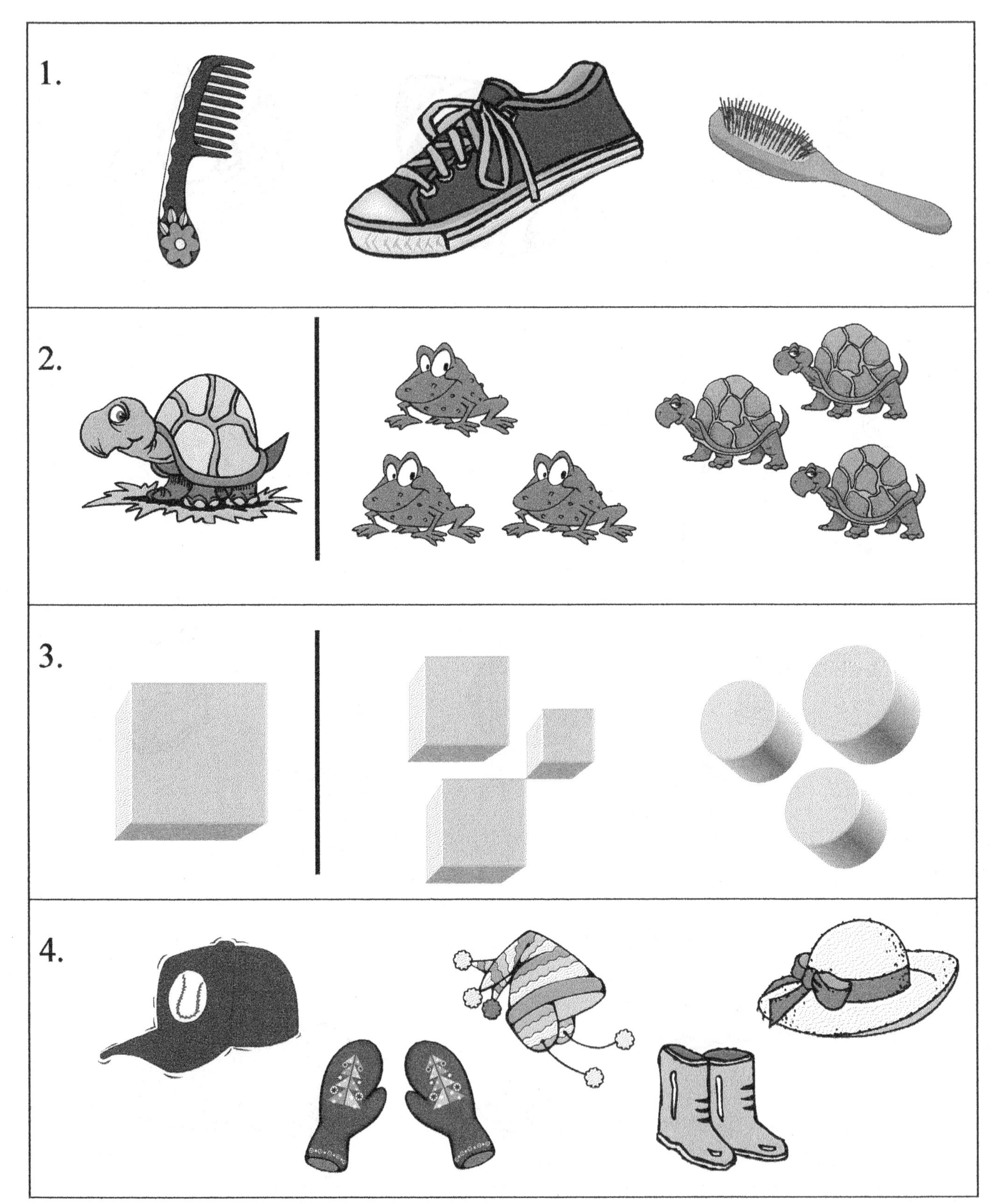

Directions:

1) Cross out the one that doesn't belong.
2 - 3) Circle the group where the object belongs.
4) Cross out the things that a person can wear on their head.

Lesson #14

Directions:

1) Cross out the item that doesn't belong to the group of round things.
2) Circle the small rabbits in red.
3) Cross out the item that doesn't belong in a group of animals with legs.
4) Circle the group where the object belongs.

Lesson #15

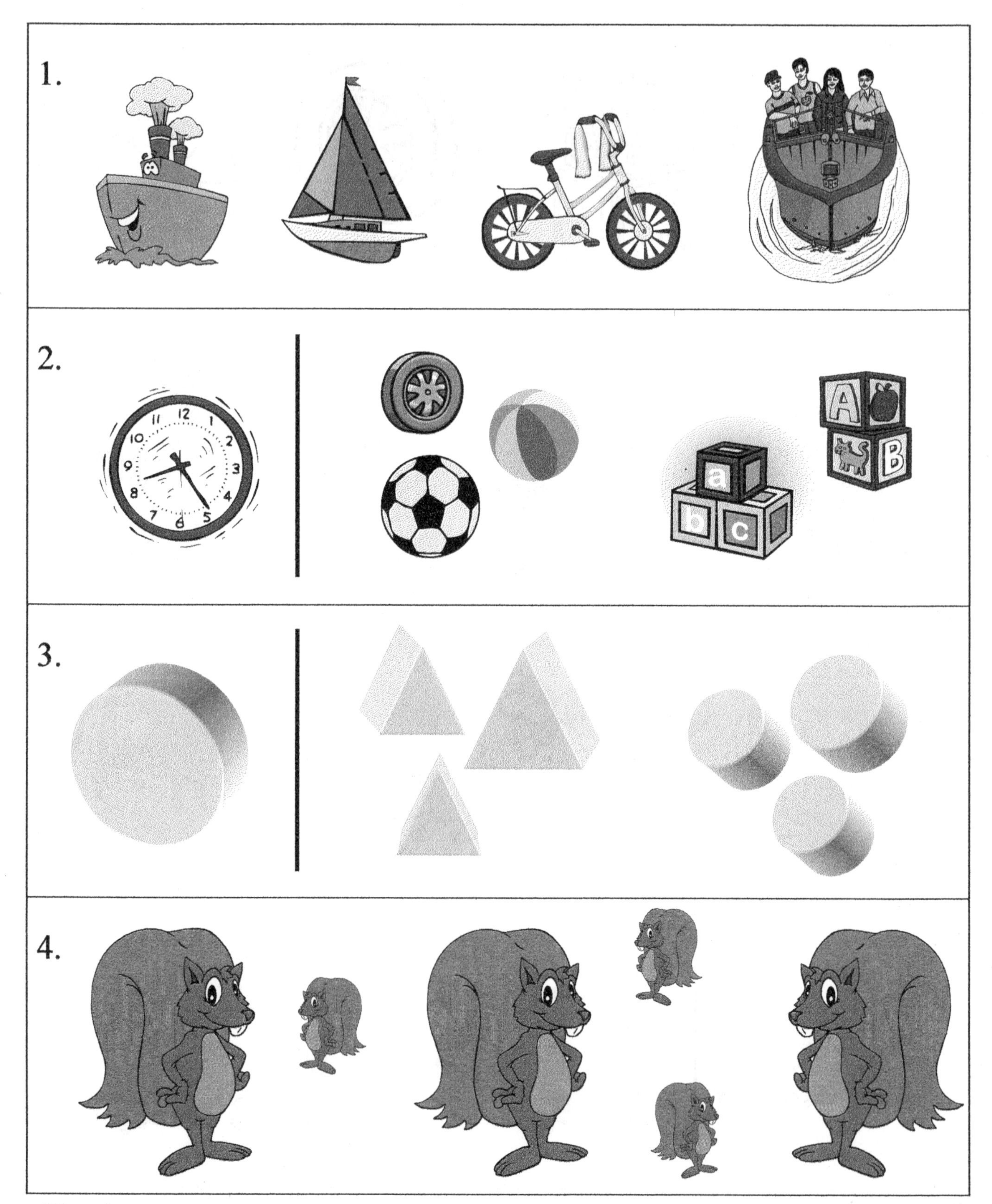

Directions:

1) Cross out the one that doesn't belong with the others.
2 – 3) Circle the group where the object belongs.
4) Cross out the big squirrels.

Lesson #16

Directions:

1) Cross out the one that doesn't belong.
2) Cross out the ones that can fly.
3 - 4) Circle the group where the object belongs.

Lesson #17

Directions:

1) Circle the object that is not tall.
2) Circle the object on the top of the tree and the one on top of the car.
3) Circle the object on top of each roof.
4) Circle the group where the object belongs.

Lesson #18

Directions:

1) Circle the hat on top of the snowman.
2) Cross out the horse that doesn't belong.
3) Cross out the things that a person can eat.
4) Color the large ducks yellow.

Lesson # 19

Directions:

1) Draw a hat on top of the snowman and one on top of the boy's head.
2) Circle the animal in the middle.
3) Cross out the one that doesn't belong.
4) Cross out the things you would wear in the winter.

Lesson #20

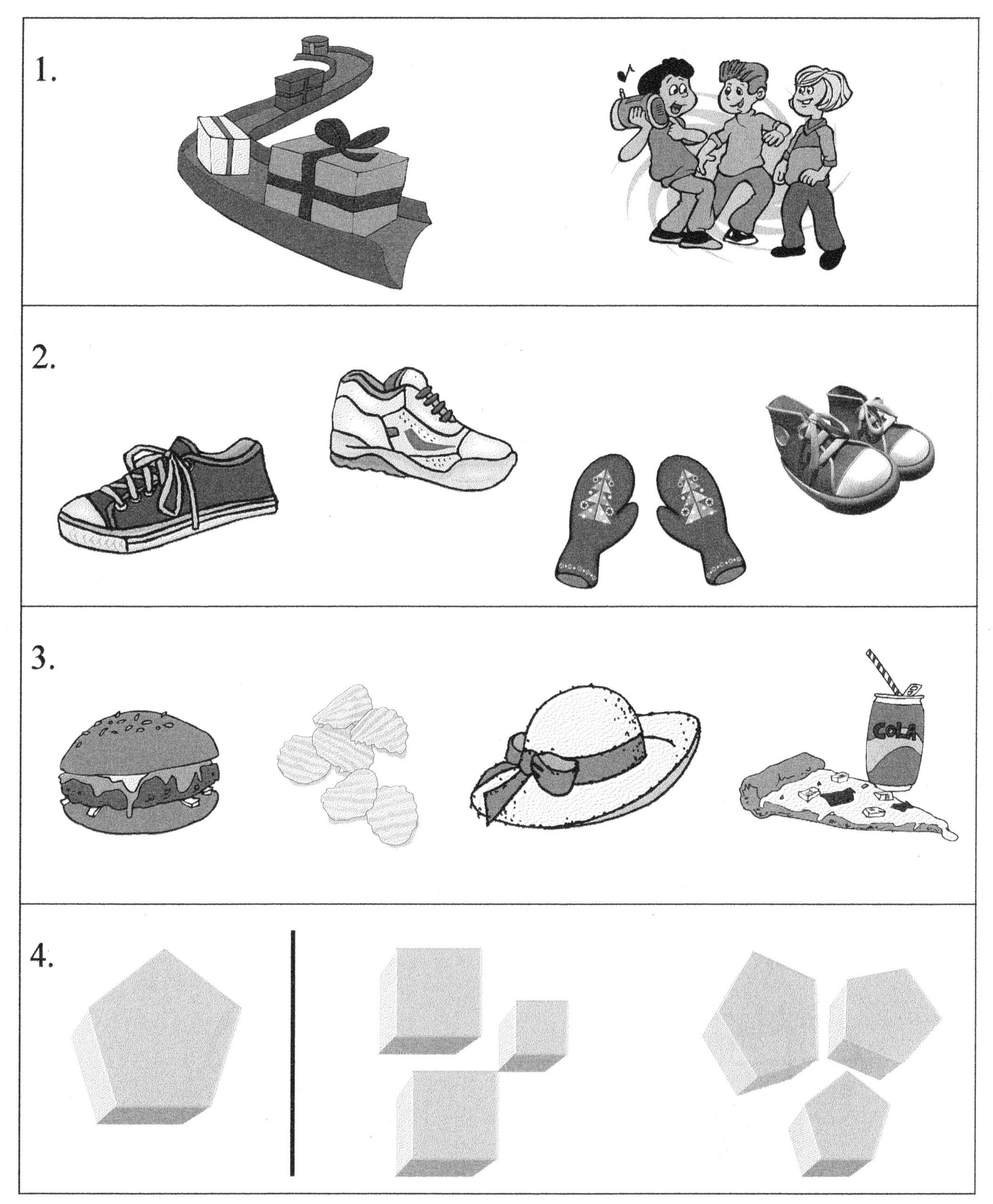

Directions:

1) Cross out the two presents in the middle. Cross out the person in the middle.
2) Circle the object that doesn't belong.
3) Cross out the object that doesn't belong.
4) Circle the group where the shape belongs.

Lesson #21

Directions:

1) Circle the group where the object belongs.
2) Color the middle circle blue. Color the bottom box red.
3) Cross out the things that people can wear.
4) Cross out what's on top of each car.

Lesson #22

Directions:

1) Circle the mittens under the table.
2) Circle the group where the shoes belong.
3) Cross out the dog in the middle.
4) Circle the group where the object belongs.

Lesson #23

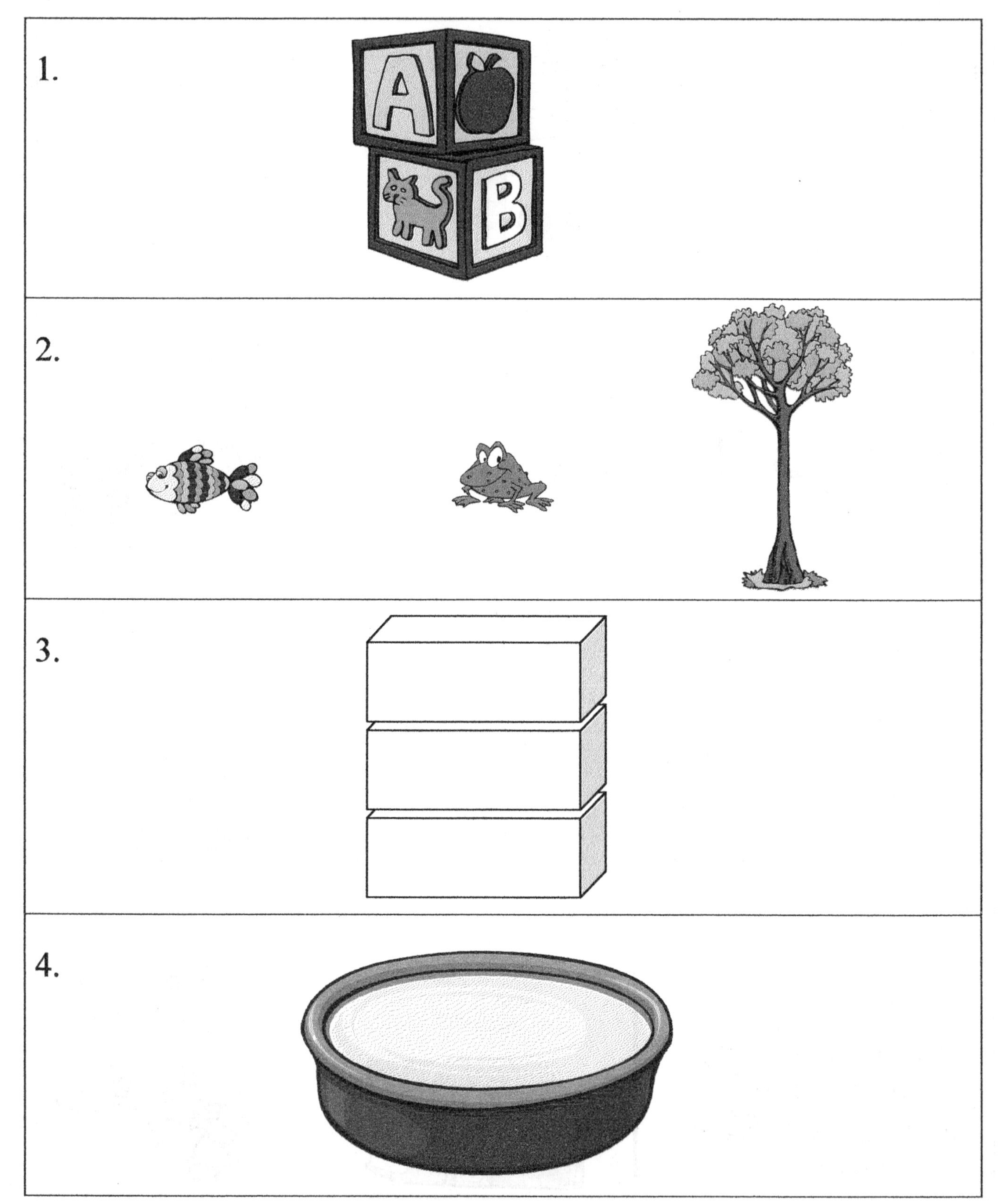

Directions:

1) Circle the block that is on top.
2) Cross out the one that is not small.
3) Color the bottom box yellow and the top box blue.
4) Draw something red inside the bowl.

Lesson #24

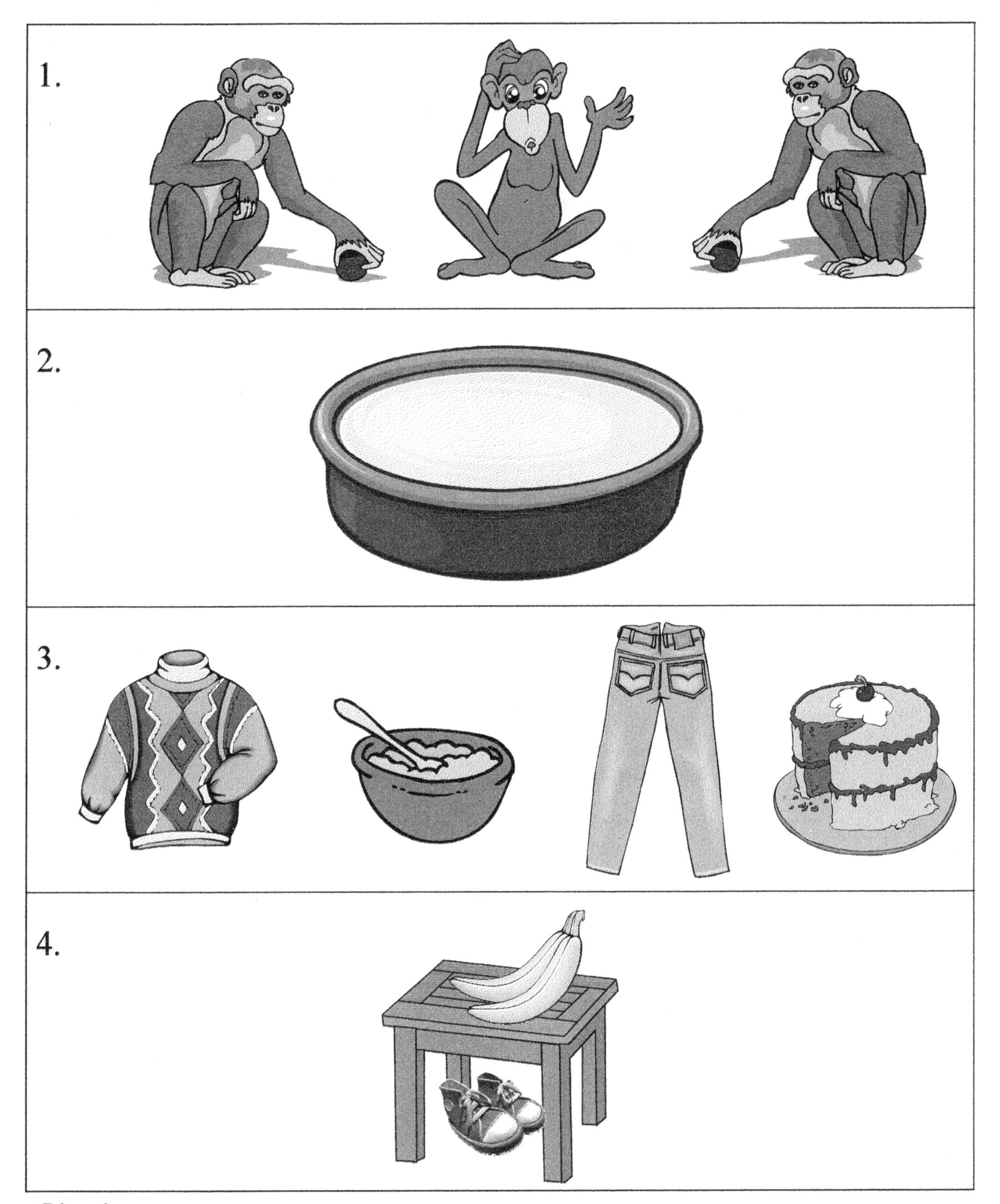

Directions:

1) Cross out the monkey that doesn't belong.
2) Draw something yellow outside of the bowl.
3) Circle the items a person can wear.
4) Circle the object under the table.

Miss Gaby

Lesson #25

1.

2.

3.

4.

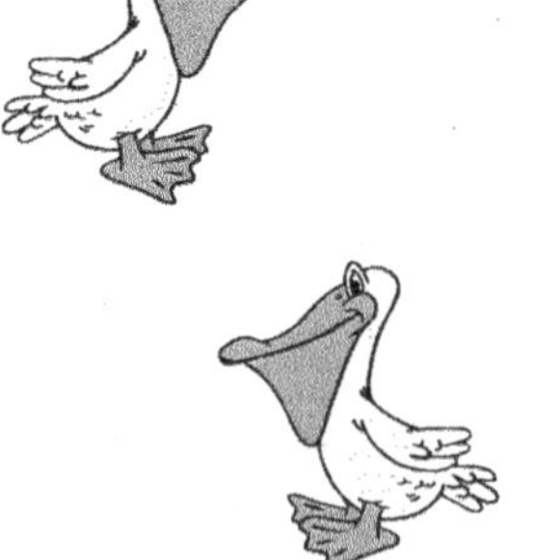

Directions:

1) Circle the animal in the middle.
2) Draw a ball on top of the chair.
3) Cross out the one that doesn't belong.
4) Circle the small objects in red.

Lesson #26

Directions:

1) Circle the animal that is before the dog.
2) Circle the animal that is after the monkey.
3) Circle the animal that is before the chicks.
4) Circle the animal in the middle.

Lesson #27

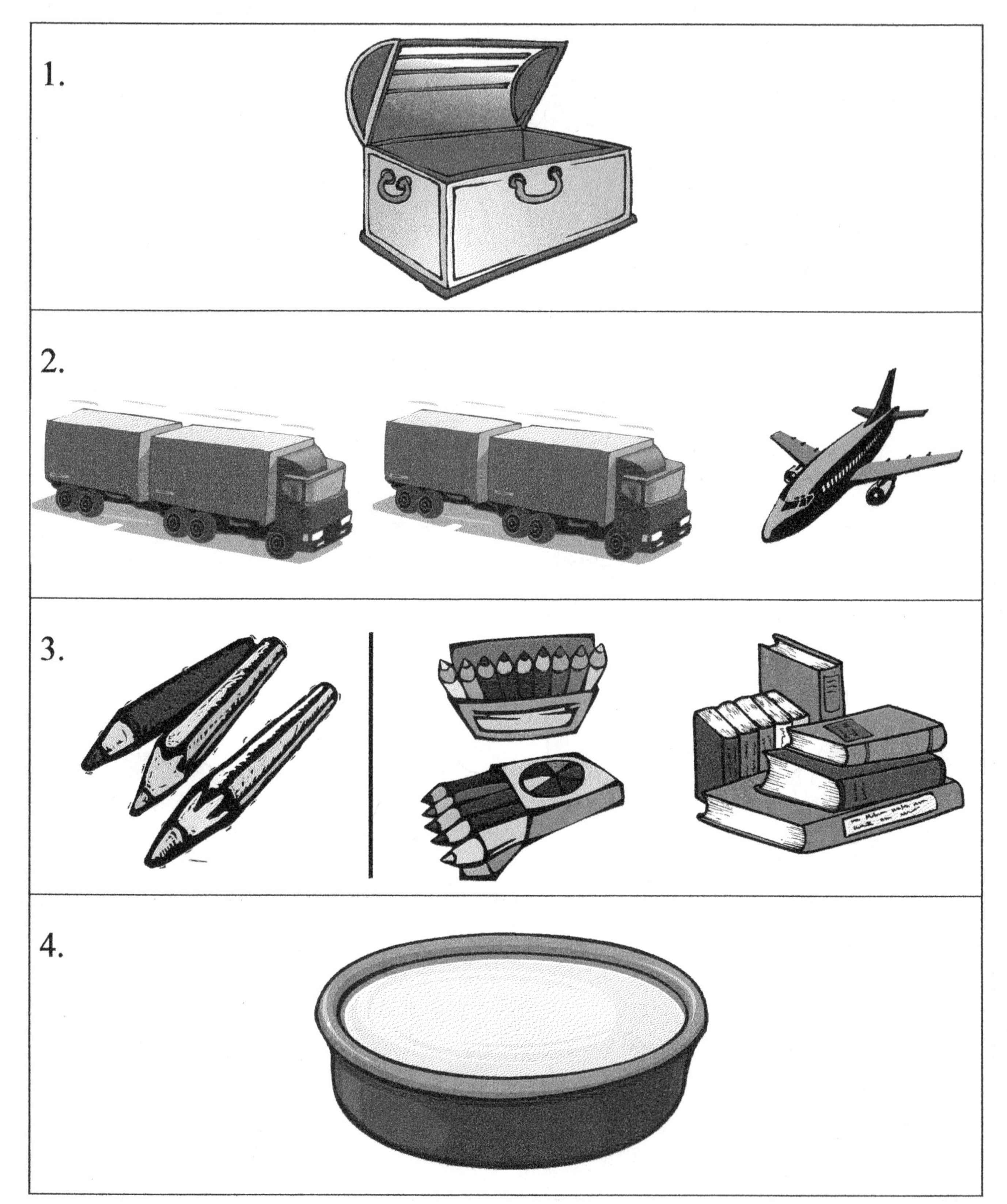

Directions:

1) Draw a toy inside the box.
2) Cross out the one that doesn't belong.
3) Circle the group where the object belongs.
4) Color the inside of the bowl blue.

Lesson #28

Directions:

1) Cross out the object between the shoes.
2) Draw an apple on top of the table.
3) Cross out the picture after the bicycle.
4) Draw a banana next to the monkey.

Lesson #29

Directions:

1) Circle the animal before the turkey.
2) Cross out the object that doesn't belong.
3) Draw water under the boats.
4) Circle the object between the houses.

Lesson #30

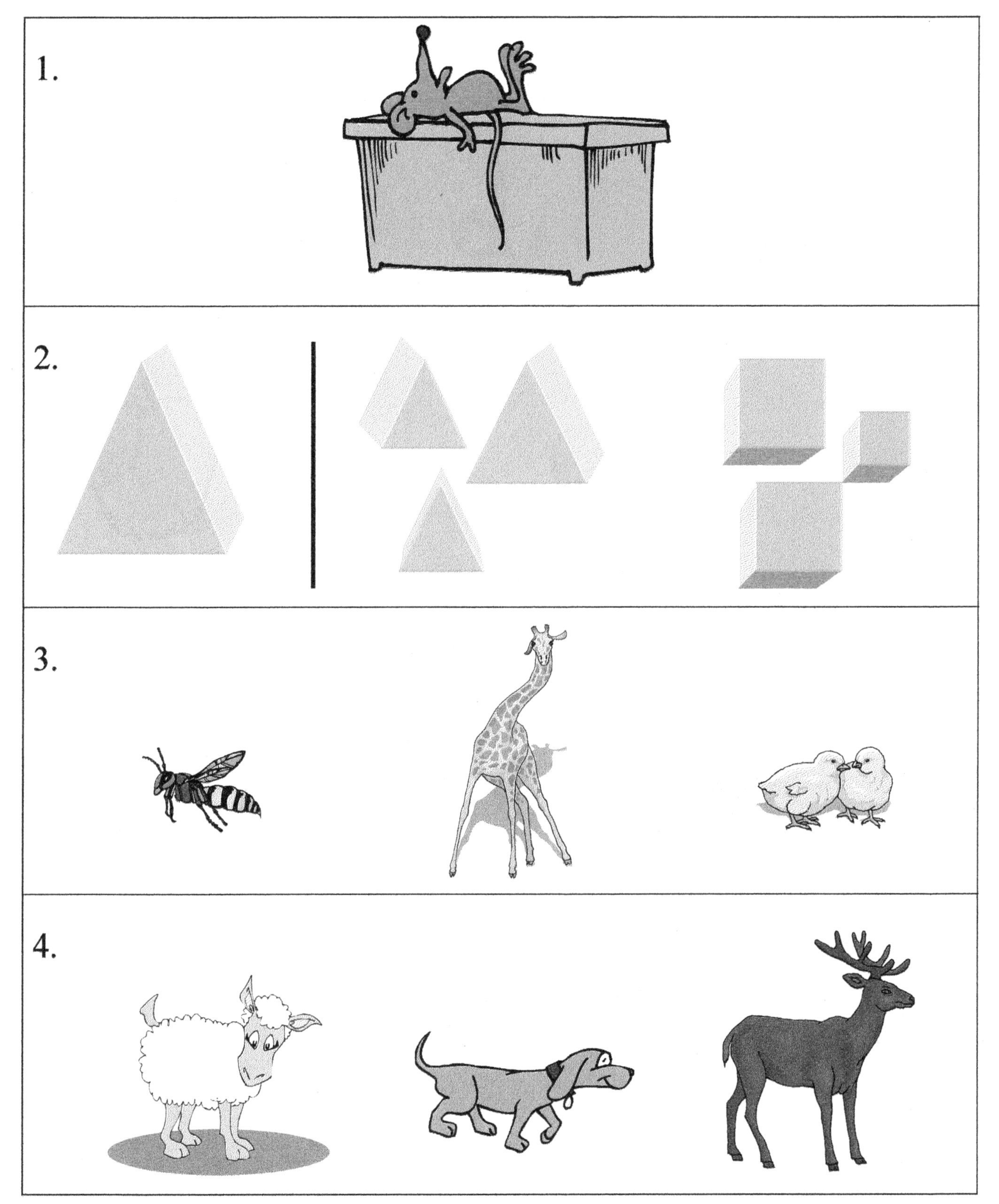

Directions:

1) Circle the object on top of the table.
2) Circle the group where the shape belongs.
3) Cross out the one that doesn't belong.
4) Cross out the animal after the sheep.

Lesson #31

Directions:

1) Circle the apples under the tree.
2) Cross out the food that doesn't belong.
3) Circle the object next to the house.
4) Circle the group where the animal belongs.

Lesson #32

Directions:

1) Circle the food after the pie.
2) Cross out the object under the table.
3) Circle the object above the house.
4) Cross out the things a person can ride.

Lesson #33

Directions:

1) Circle the picture before the snowman.
2) Circle the group where the object belongs.
3) Cross out the picture before the fries.
4) Cross out the one that doesn't belong.

Lesson #34

Directions:

1) Circle the object under the cake. Circle the object on top of the pie.
2) Cross out the things a person can eat.
3) Circle the object on top of the table.
4) Cross out the picture before the boat.

Lesson #35

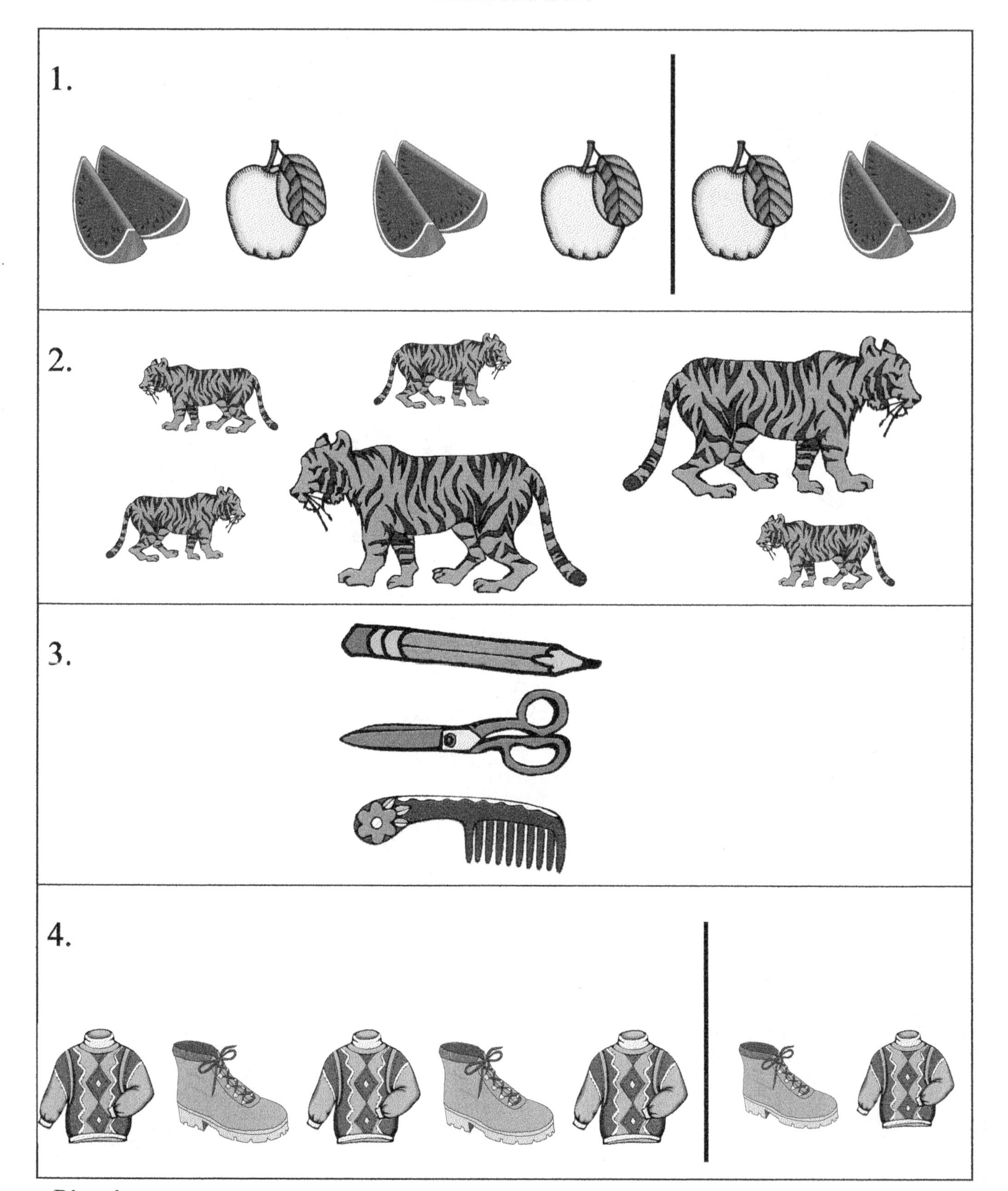

Directions:

1) The items before the line form a pattern. Circle the item that is likely to come next.
2) Circle the small tigers in red.
3) Circle the object below the scissors.
4) The items before the line form a pattern. Circle the item that is likely to come next.

Lesson #36

Directions:

1) Circle the picture after the squirrel.
2) The items before the line form a pattern. Circle the picture that is likely to come next.
3) Circle the bottom book.
4) Color one counter for each mouse.

Lesson #37

Directions:

1) Circle the animal that will come next in the pattern.
2) Color one counter for each shoe.
3) Circle the animal in the middle.
4) Circle the picture that doesn't belong.

Lesson #38

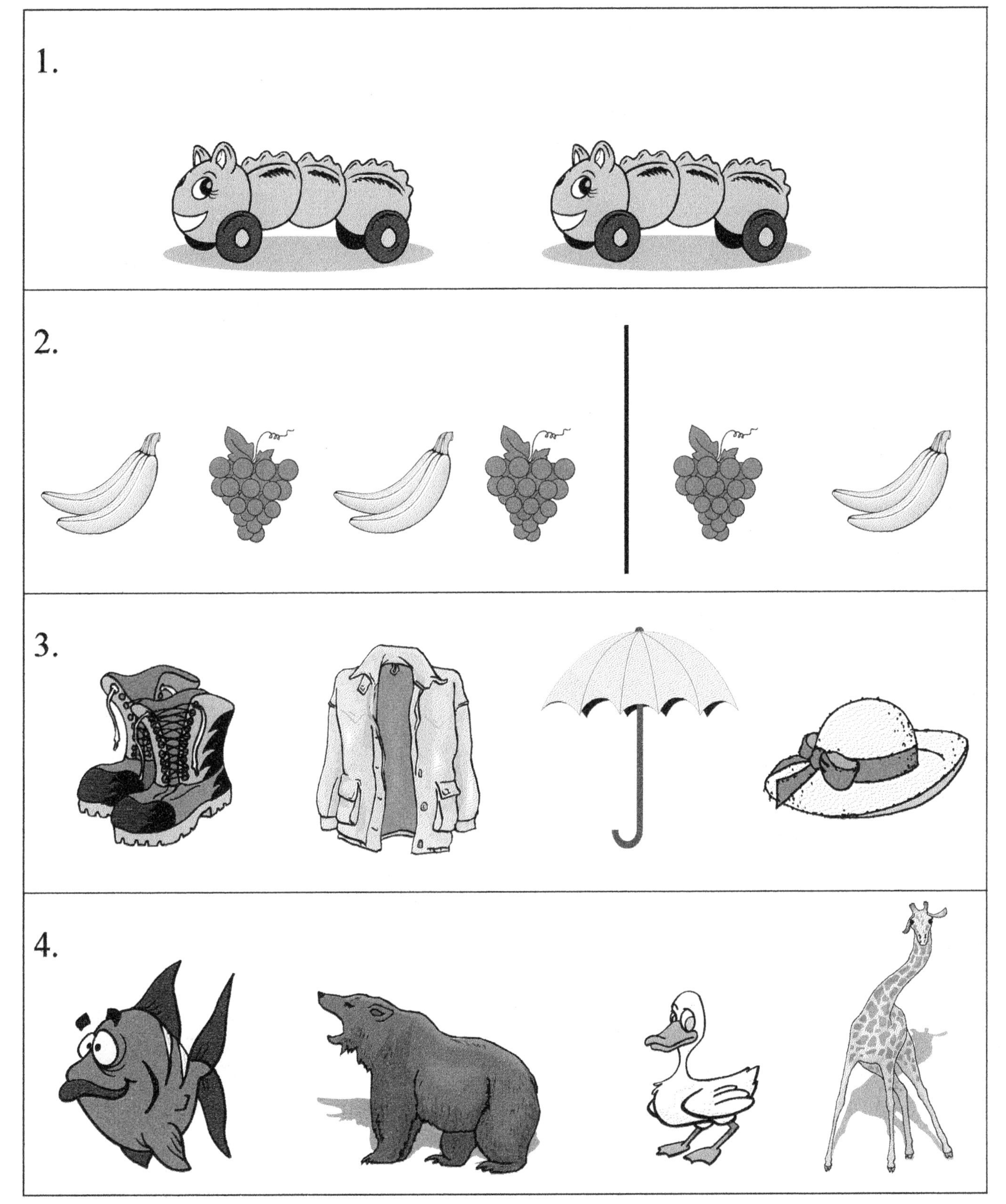

Directions:

1) Use counters to show the same number of toys. Draw the counters.
2) Circle the fruit that should come next.
3) Cross out the things a person can wear.
4) Circle the animal before the duck.

Lesson #39

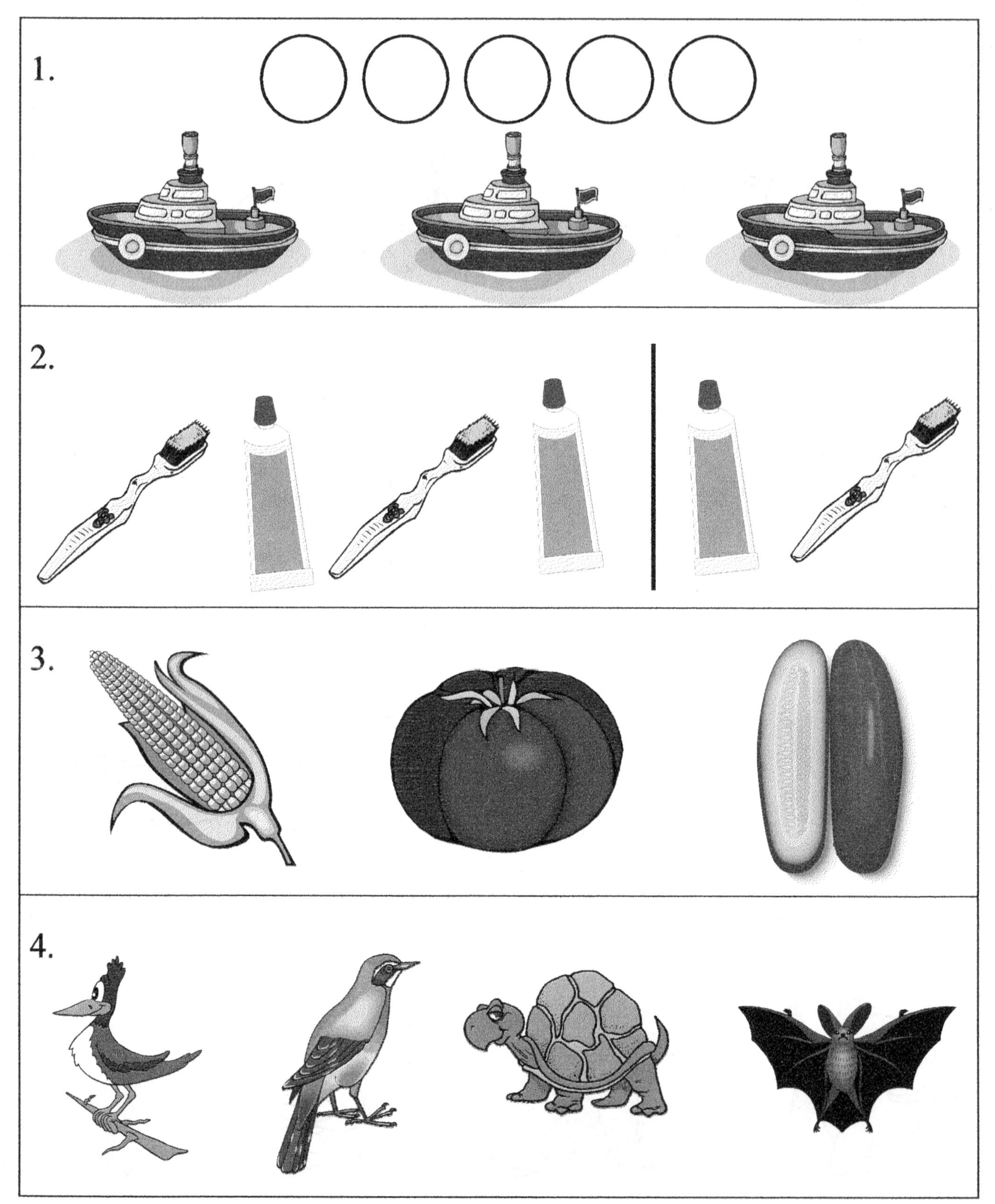

Directions:

1) Color one counter for each boat.
2) Circle the one that is likely to come next in the pattern.
3) Cross out the picture after the corn.
4) Cross out the one that doesn't belong.

Lesson #40

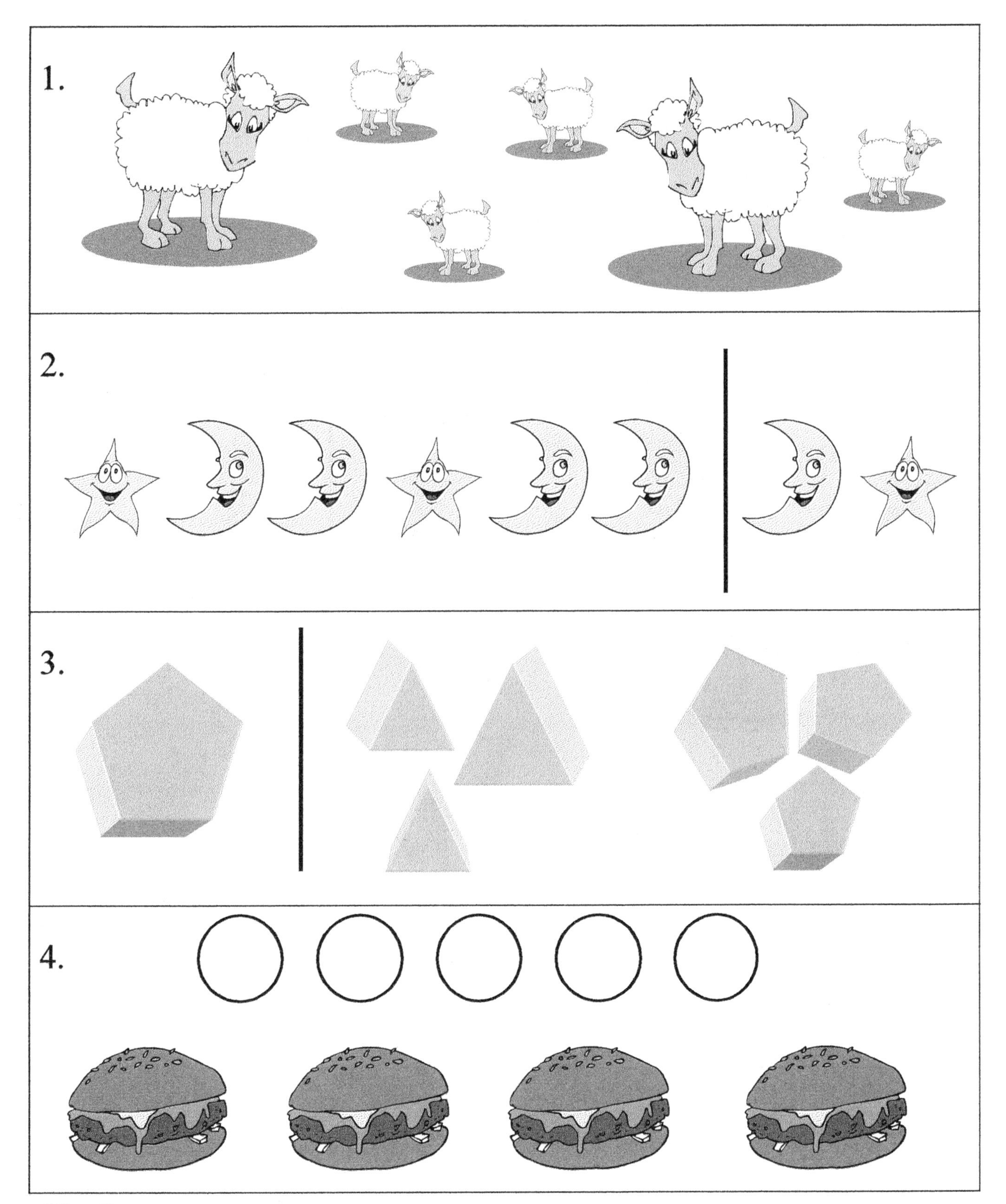

Directions:

1) Circle the small sheep yellow.
2) Circle the one that is likely to come next in the pattern.
3) Circle the group where the object belongs.
4) Color one counter for each hamburger.

Lesson #41

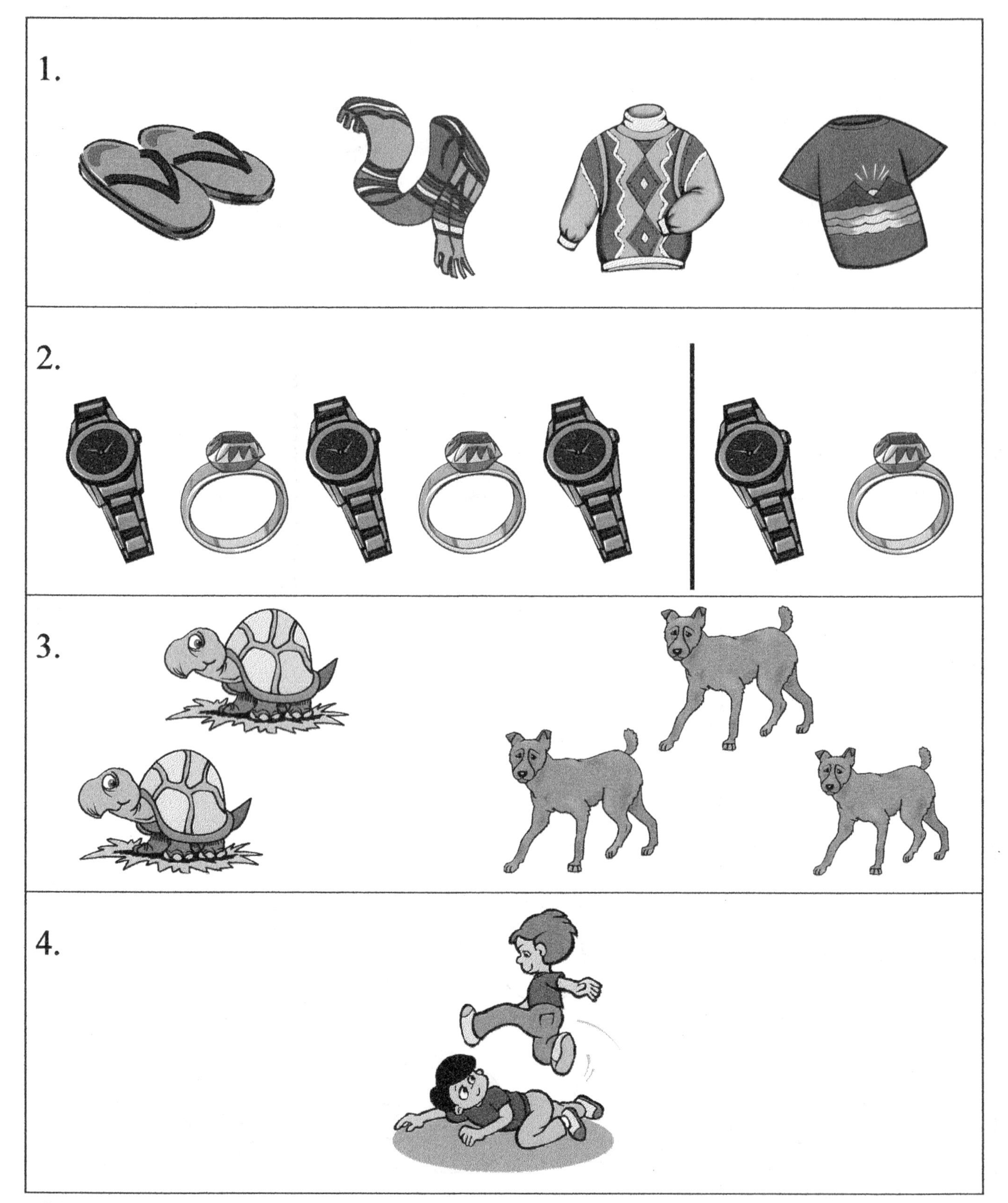

Directions:

1) Cross out the things you wear in the winter.
2) Circle the one that comes next in the pattern.
3) Count the number in each group. Circle the group that has more.
4) Cross out the boy on the bottom.

Lesson #42

Directions:

1) Circle the group where the animal belongs.
2) Cross out the object in front of the couch.
3) Circle the animal that is likely to come next in the pattern.
4) Count the number in each group. Circle the group that has more.

Lesson #43

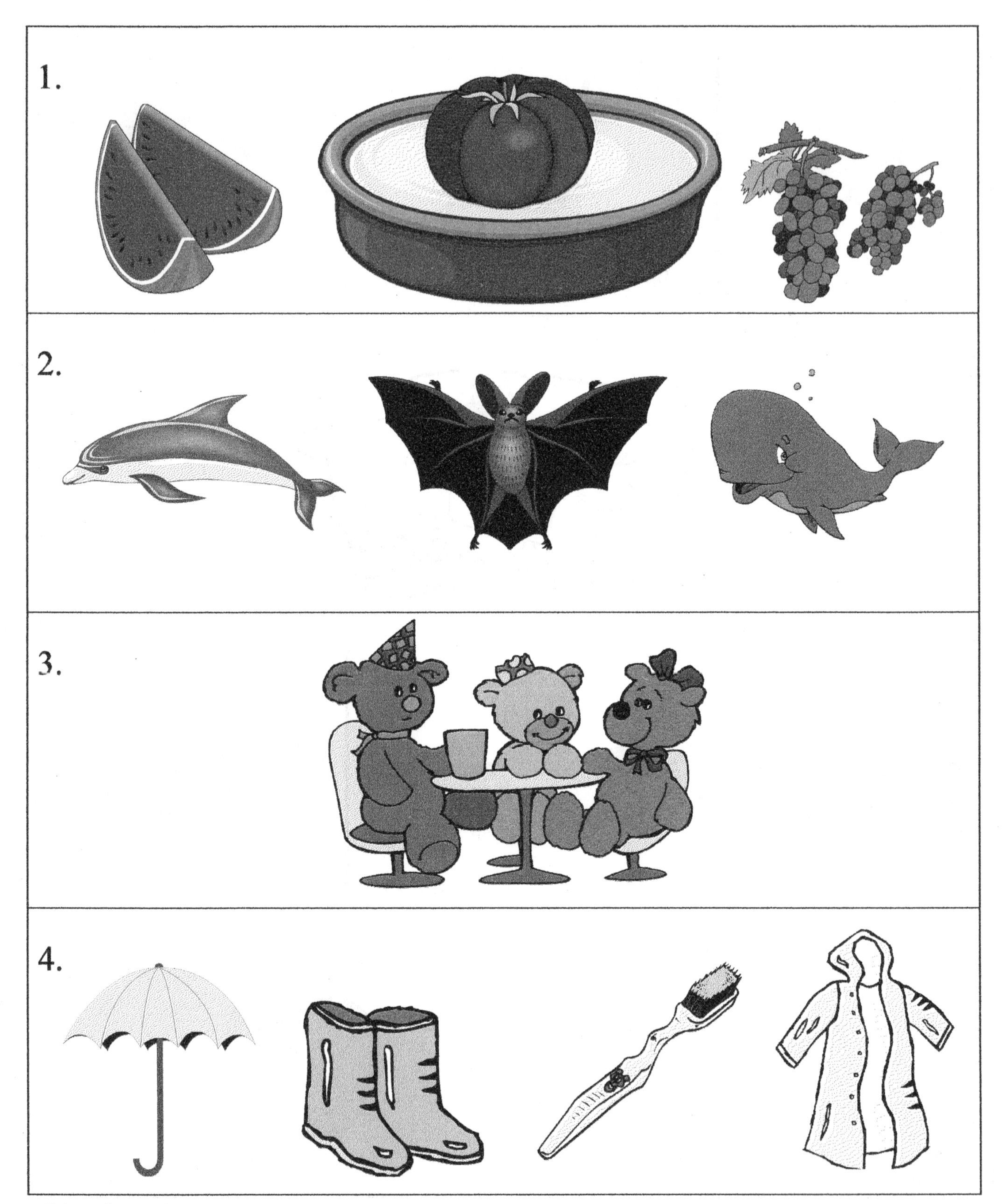

Directions:

1) Cross out the object inside the bowl.
2) Cross out the animal that doesn't belong.
3) Circle the bear in the middle.
4) Cross out the things a person can't wear.

Lesson #44

1.

2.

3.

4.

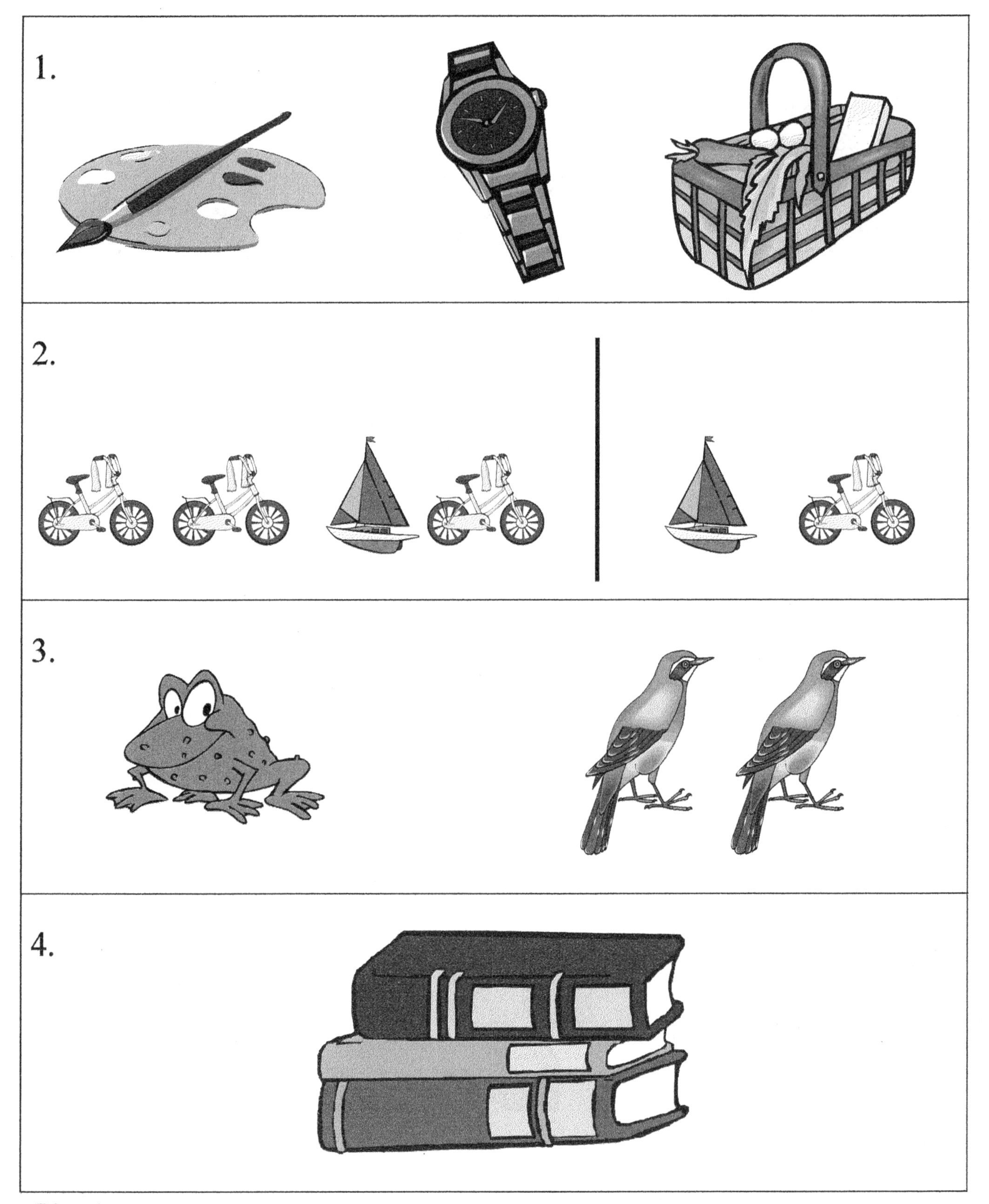

Directions:

1) Circle the picture before the watch.
2) Circle the one that is likely to come next in the pattern.
3) Count the number in each group. Circle the group that has more.
4) Cross out the book on the bottom.

Lesson #45

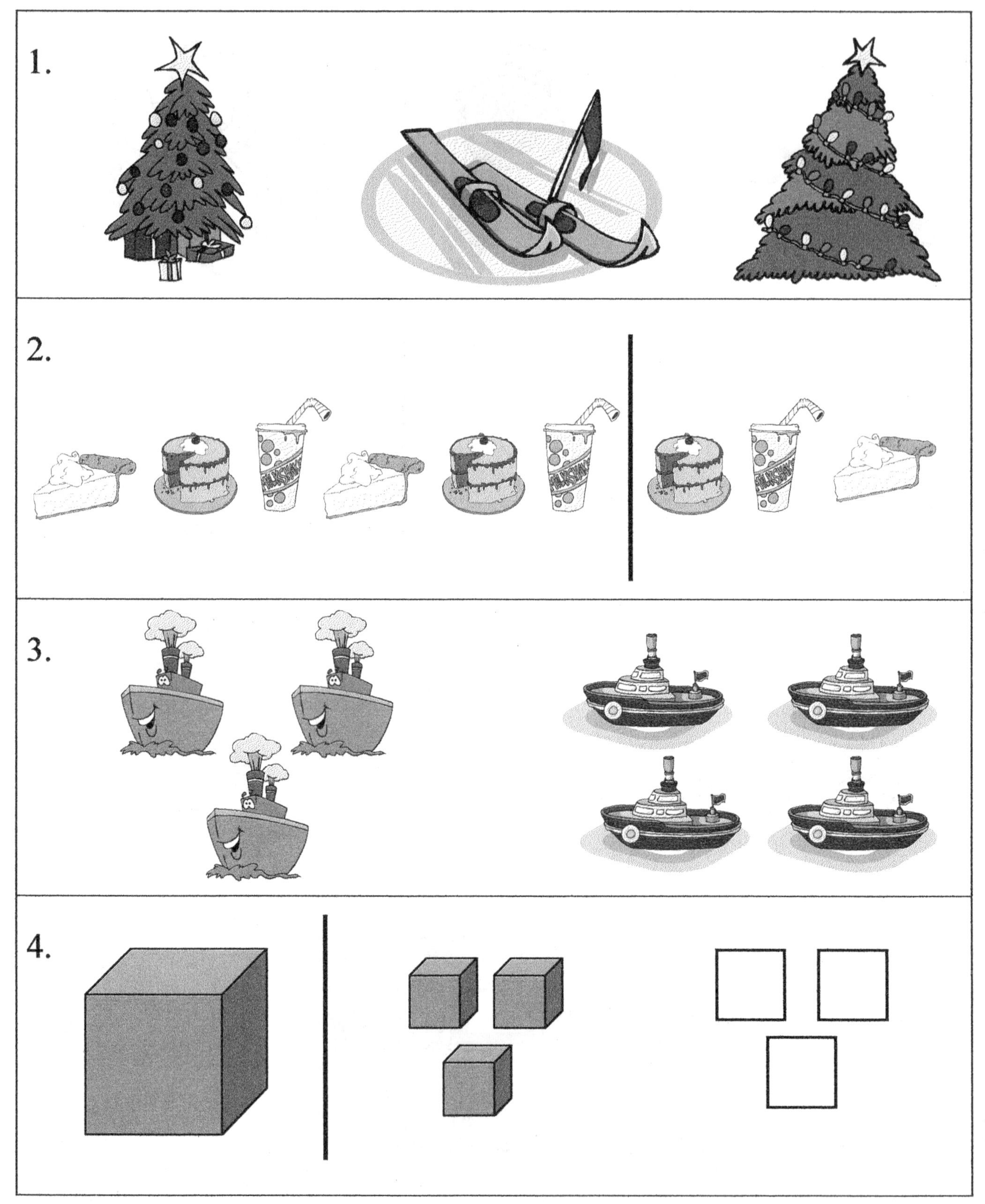

Directions:

1) Circle the object between the trees.
2) Circle the one that should come next in the pattern.
3) Count the boats in each group. Circle the group that has fewer.
4) Circle the group where the shape belongs.

Lesson #46

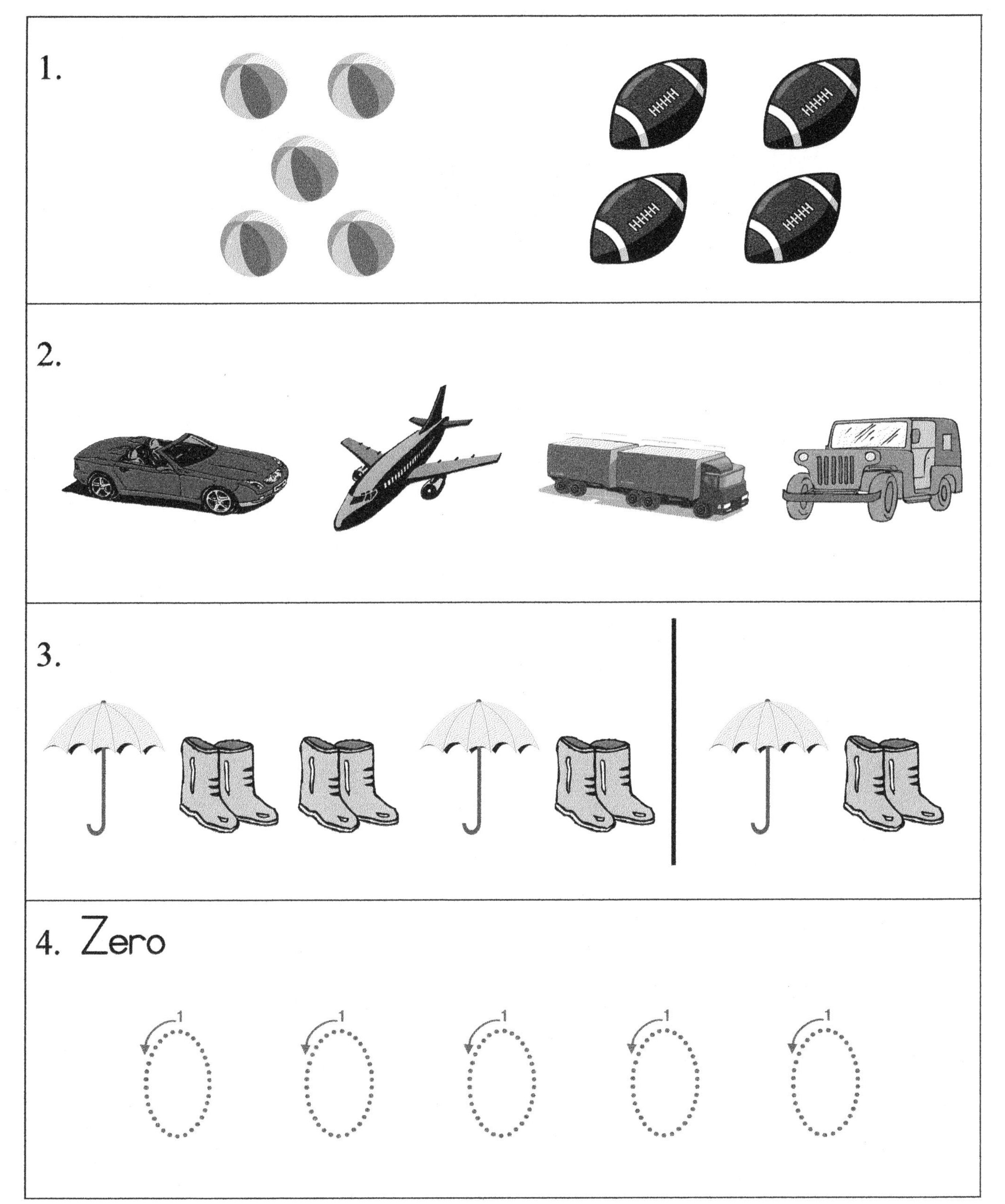

Directions:

1) Circle the group that has fewer.
2) Cross out the picture that doesn't belong.
3) Circle the one that should come next in the pattern.
4) Trace the number.

Lesson #47

Directions:

1) Cross out the animal after the bat.
2) Cross out the food that doesn't belong.
3) Trace the number.
4) Circle the group where the object belongs.

Lesson #48

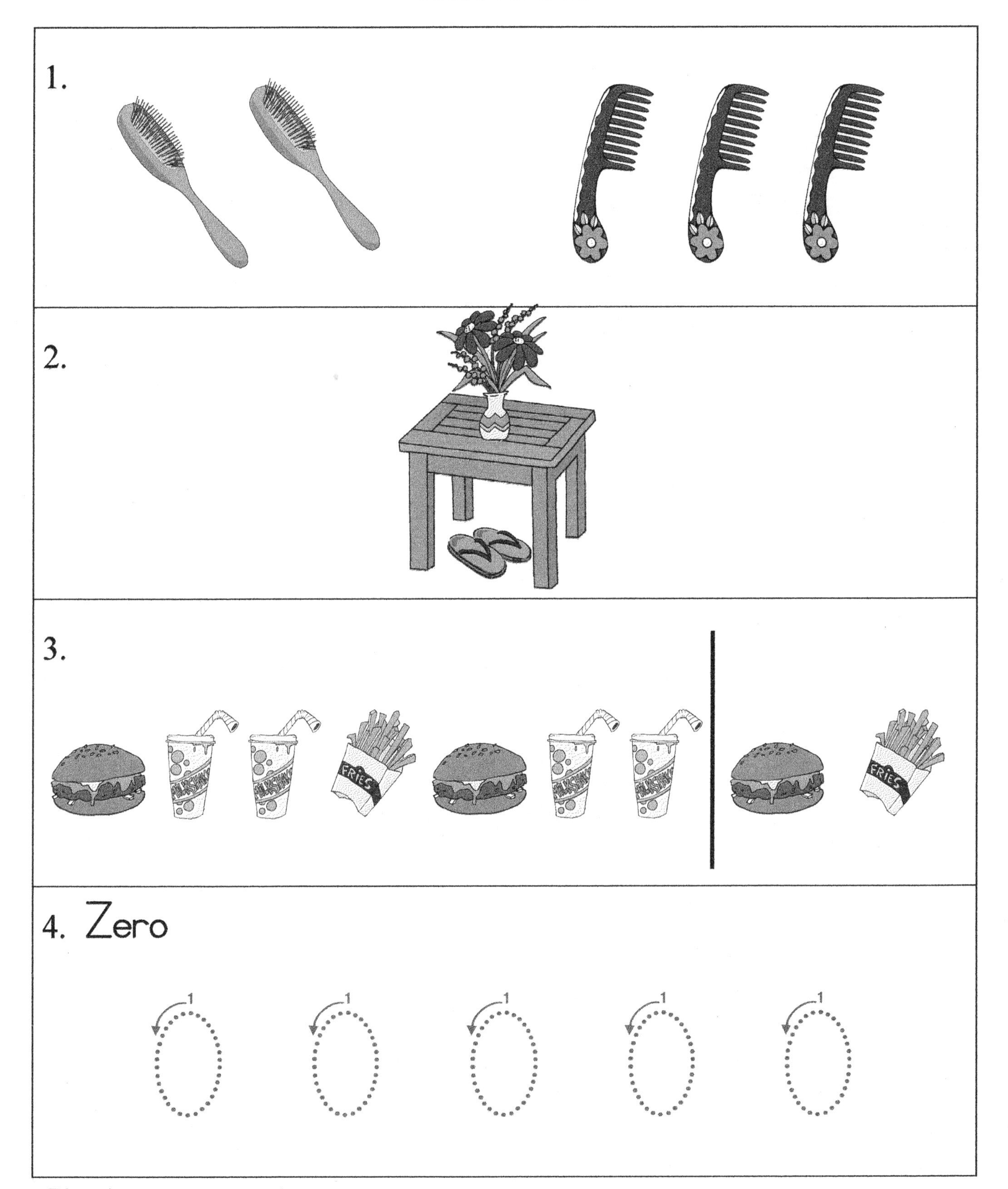

Directions:

1) Count the number in each group. Circle the group that has more.
2) Circle the item under the table.
3) Circle the object that should come next.
4) Trace the number.

Lesson #49

Directions:

1) Cross out the animal that doesn't belong.
2) Trace the number.
3) Circle the object above the boy.
4) Count the number in each group. Circle the group that has fewer.

Lesson #50

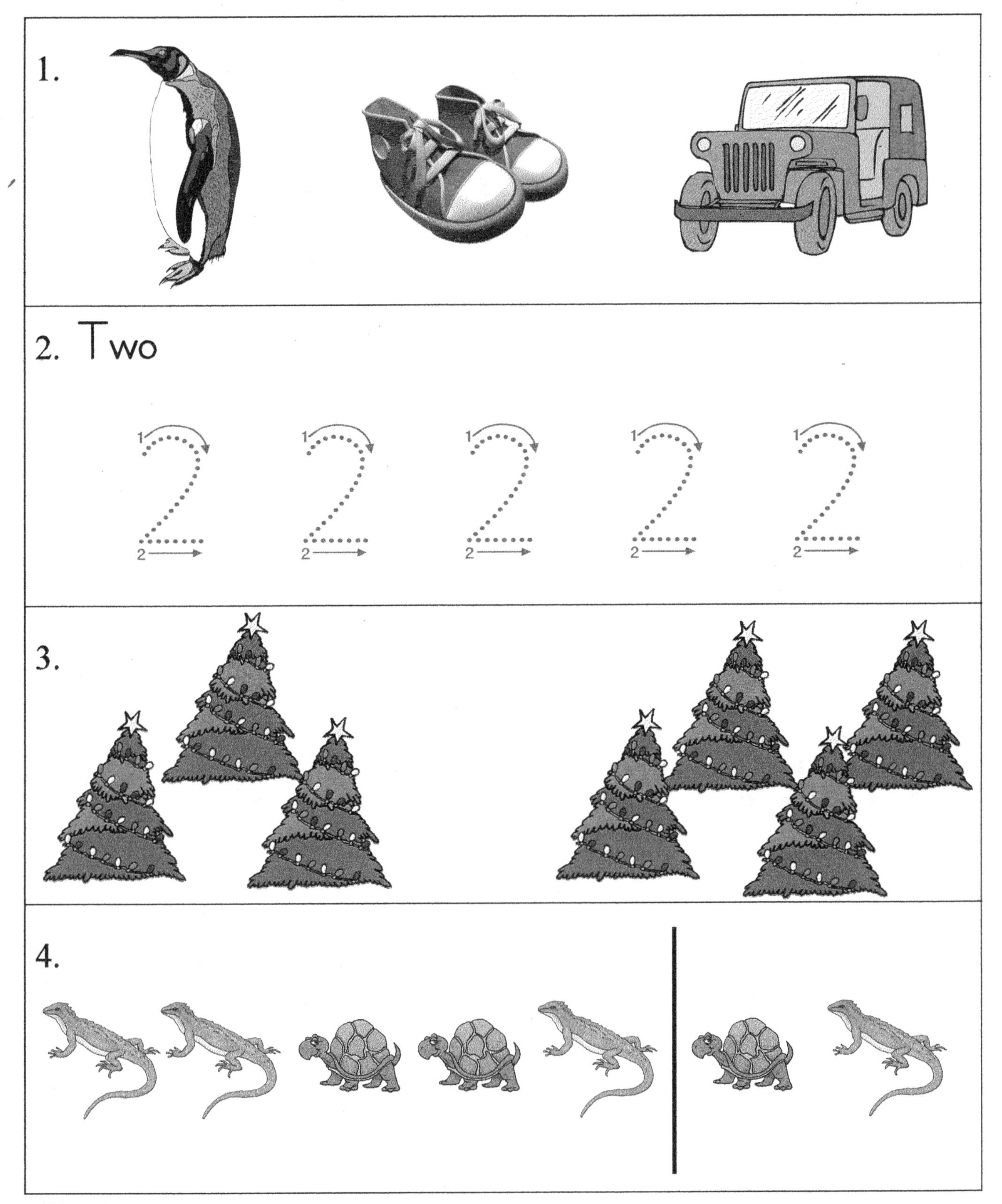

Directions:

1) Cross out the picture before the shoes.
2) Trace the number.
3) Count the number of trees in each group. Circle the group that has more.
4) Circle the animal that should come next in the pattern.

Lesson #51

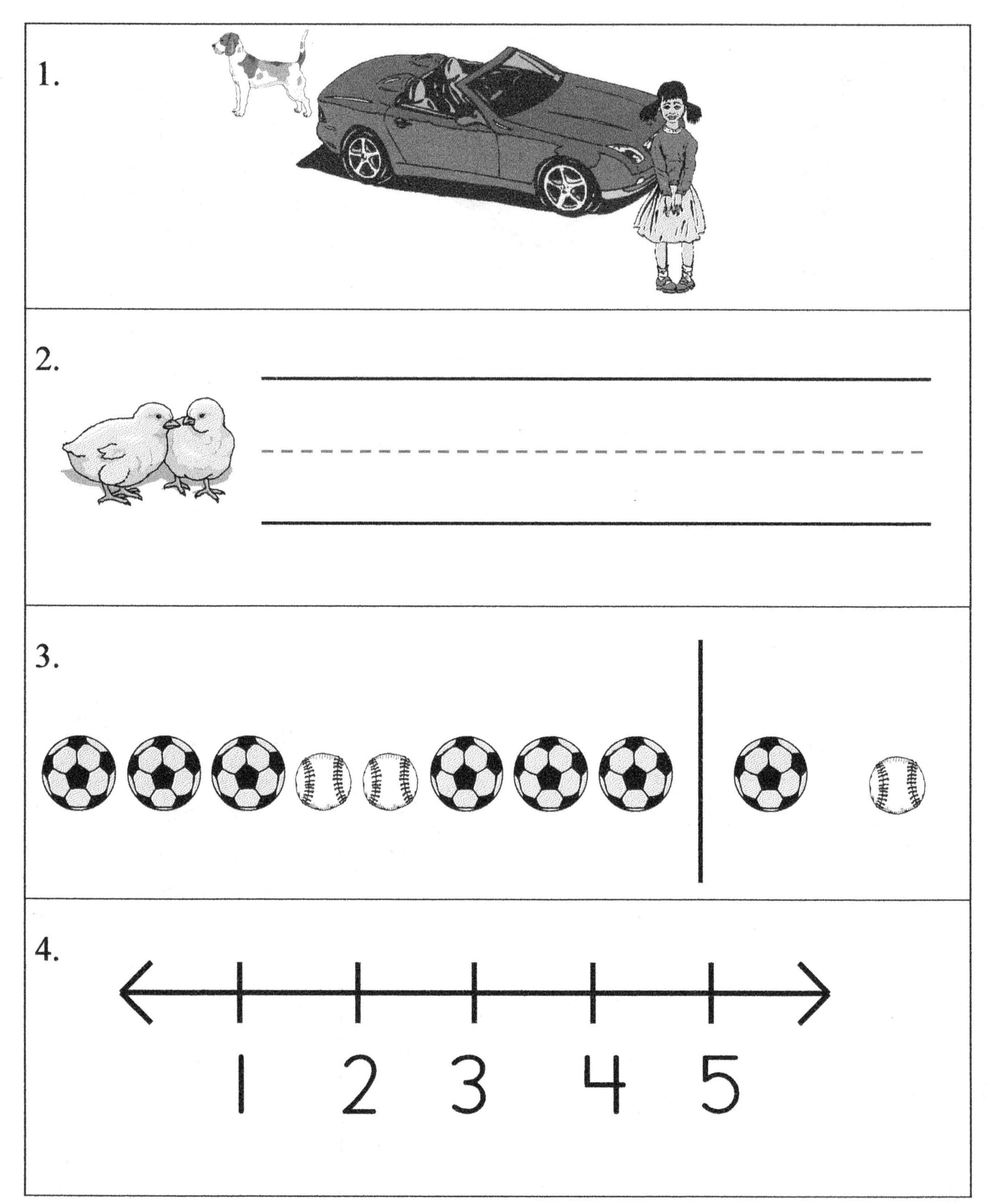

Directions:

1) Circle the object in front of the car.
2) Count the chicks. Write the number three times.
3) Circle the ball that is likely to come next in the pattern.
4) Circle the number that comes after 3.

Lesson #52

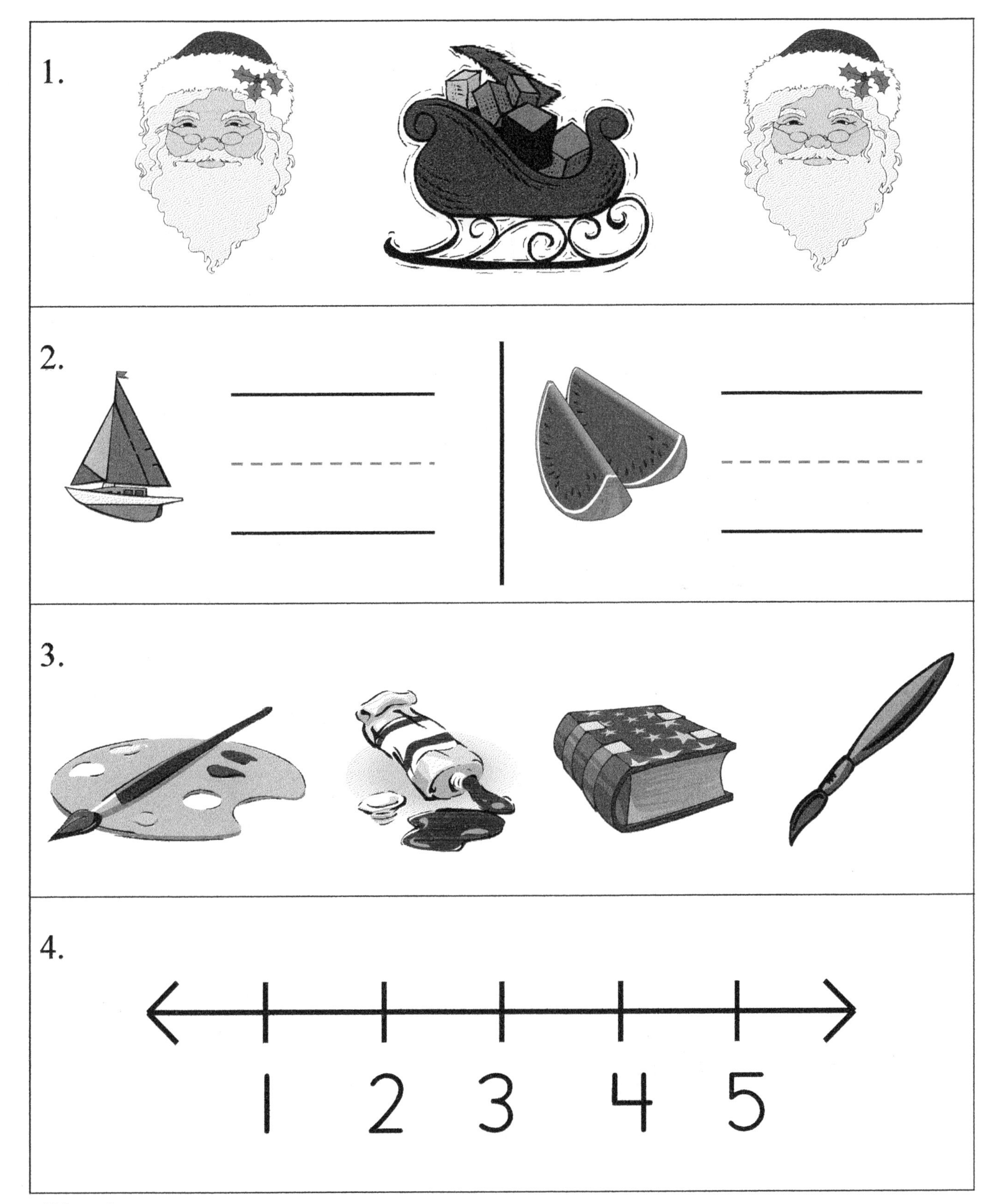

Directions:

1) Cross out the object between the Santa faces.
2) Count the boats. Write the number. How many slices of watermelon? Write the number.
3) Cross out the picture that doesn't belong.
4) Circle the number that comes after 4.

Lesson #53

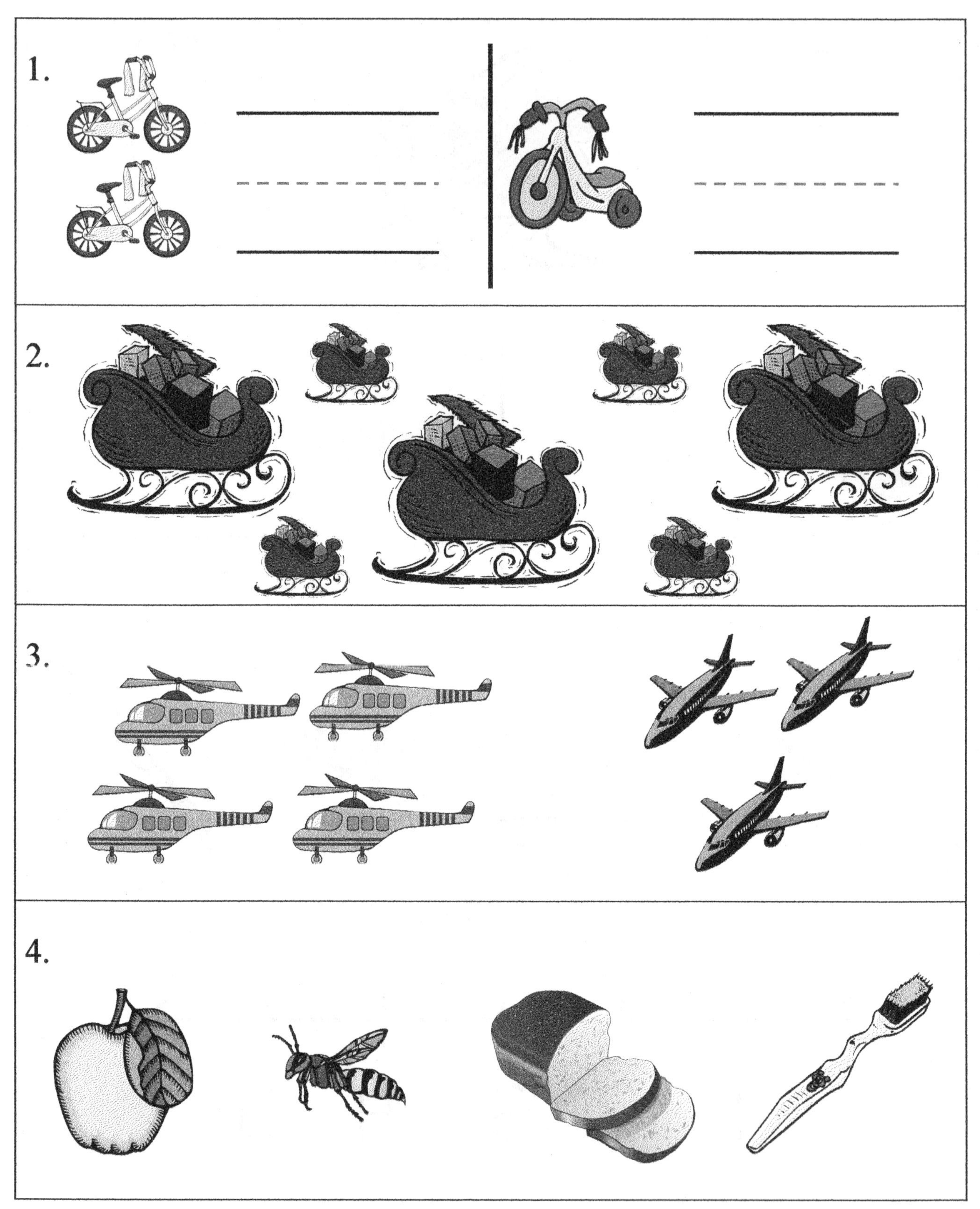

Directions:

1) Count the number of bikes in each group. Write the number.
2) Cross out the large sleighs.
3) Count the number in each group. Circle the group with more.
4) Cross out the object after the bee.

Lesson #54

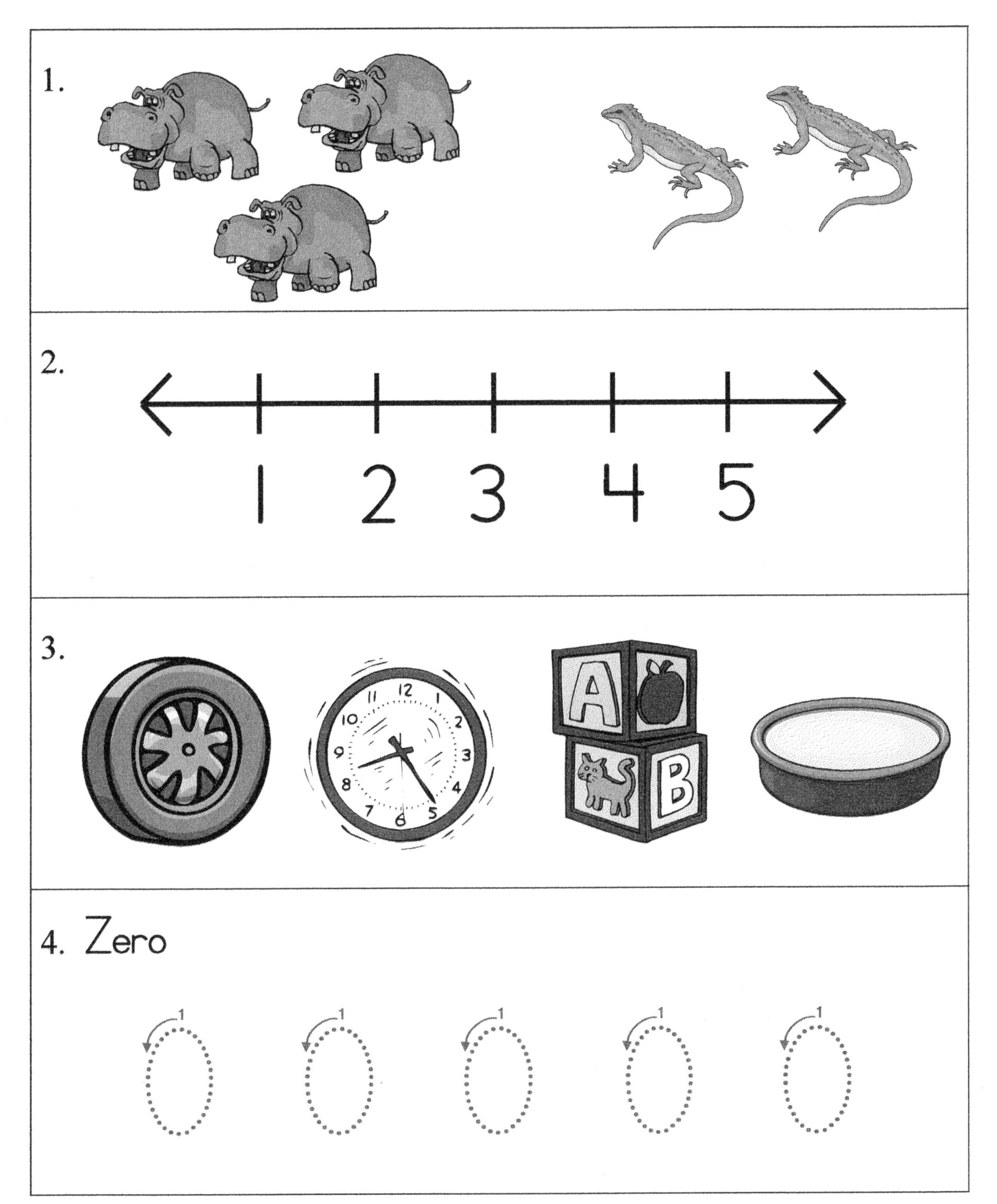

Directions:

1) Circle the group that shows 3.
2) Circle the number that comes after 1.
3) Cross out the picture that doesn't belong.
4) Trace the number.

Lesson #55

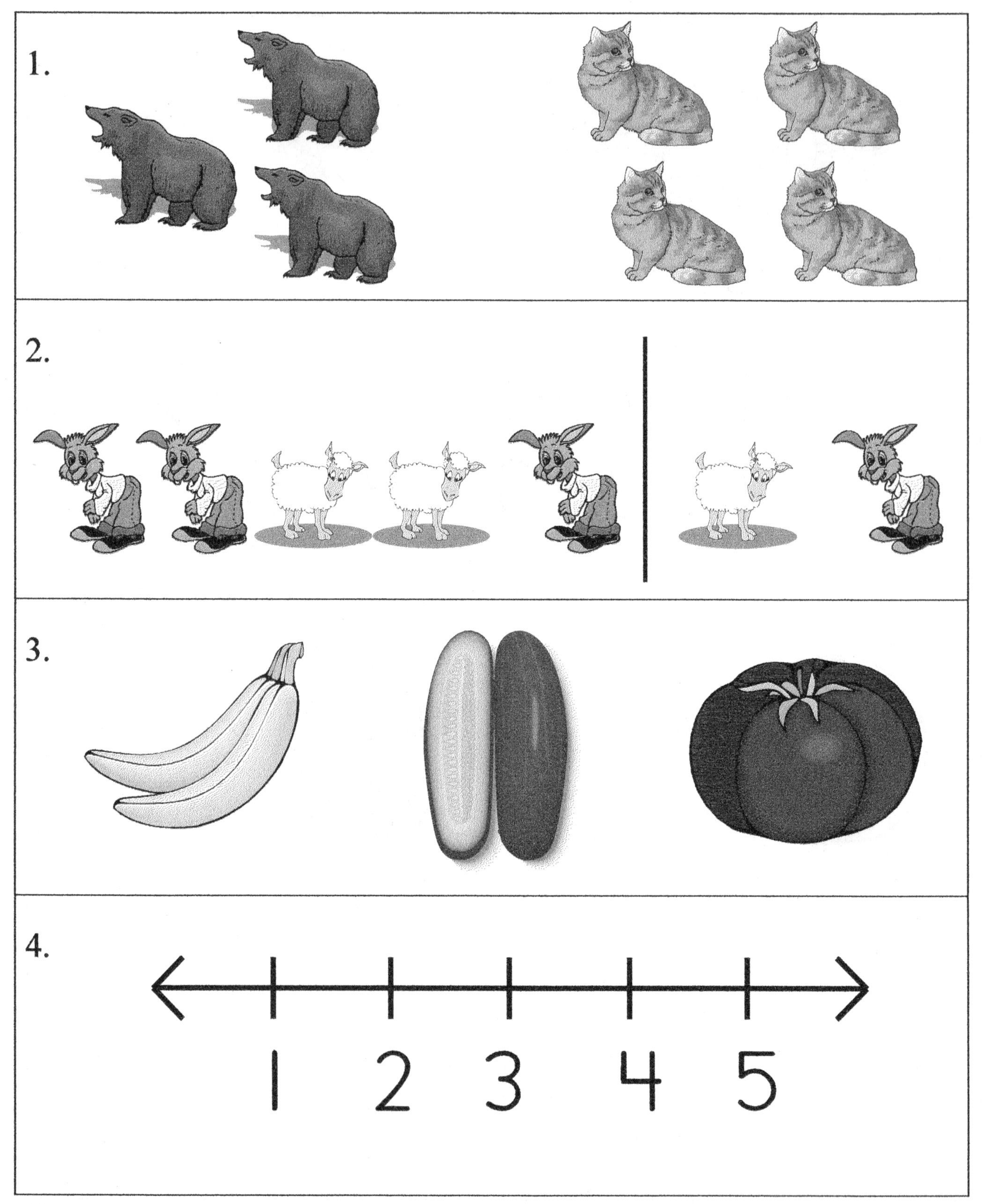

Directions:

1) Count the number in each group. Circle the group that has fewer.
2) Circle the one that should come next.
3) Cross out the picture before the tomato.
4) Circle the number that comes after 2.

Lesson #56

1.

2.

3.

4.

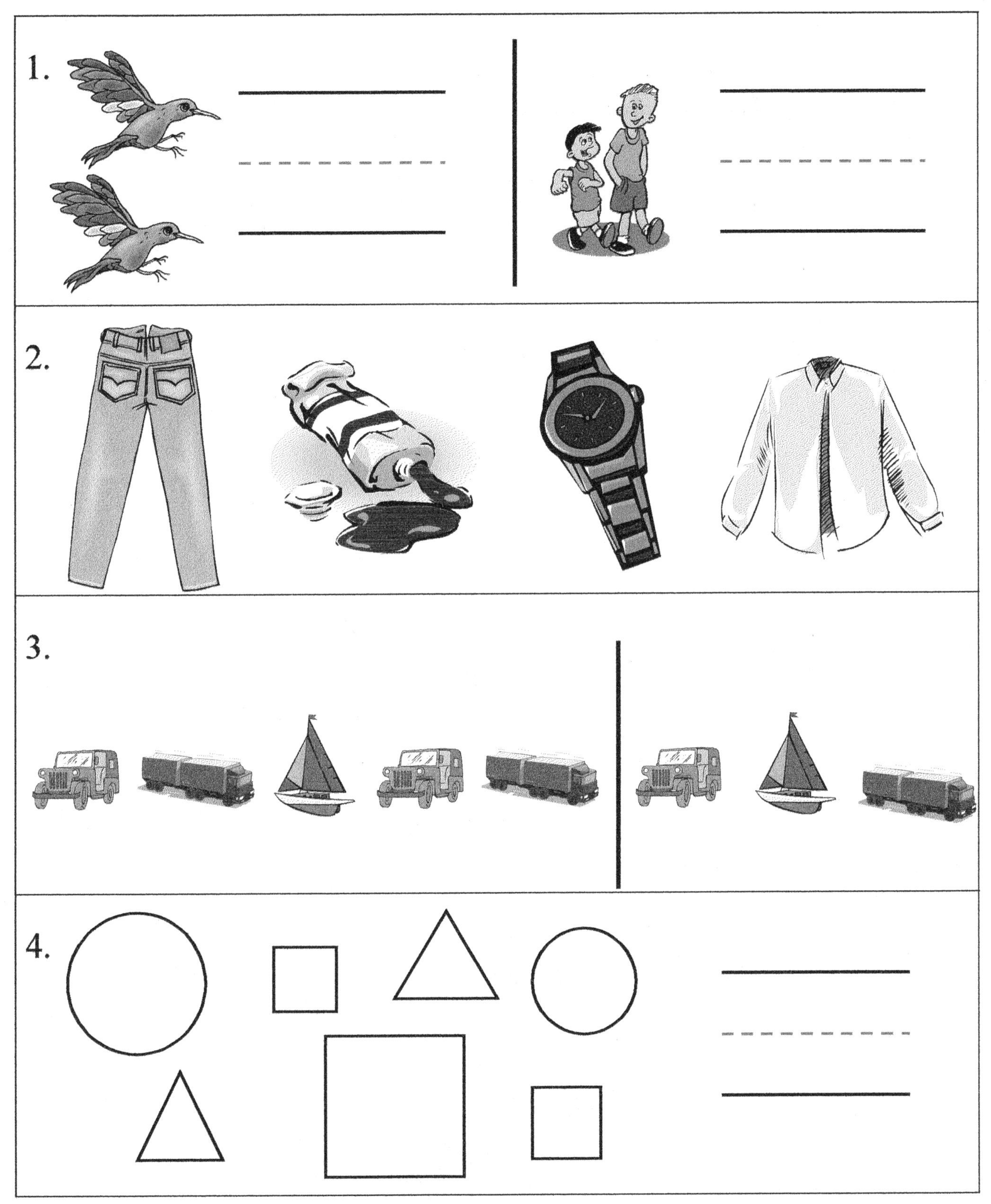

Directions:

1) Count the number in each group. Write the number.
2) Cross out the things a person can't wear.
3) Circle the one that should come next.
4) Color the circles red. Count the circles. Write the number.

Lesson #57

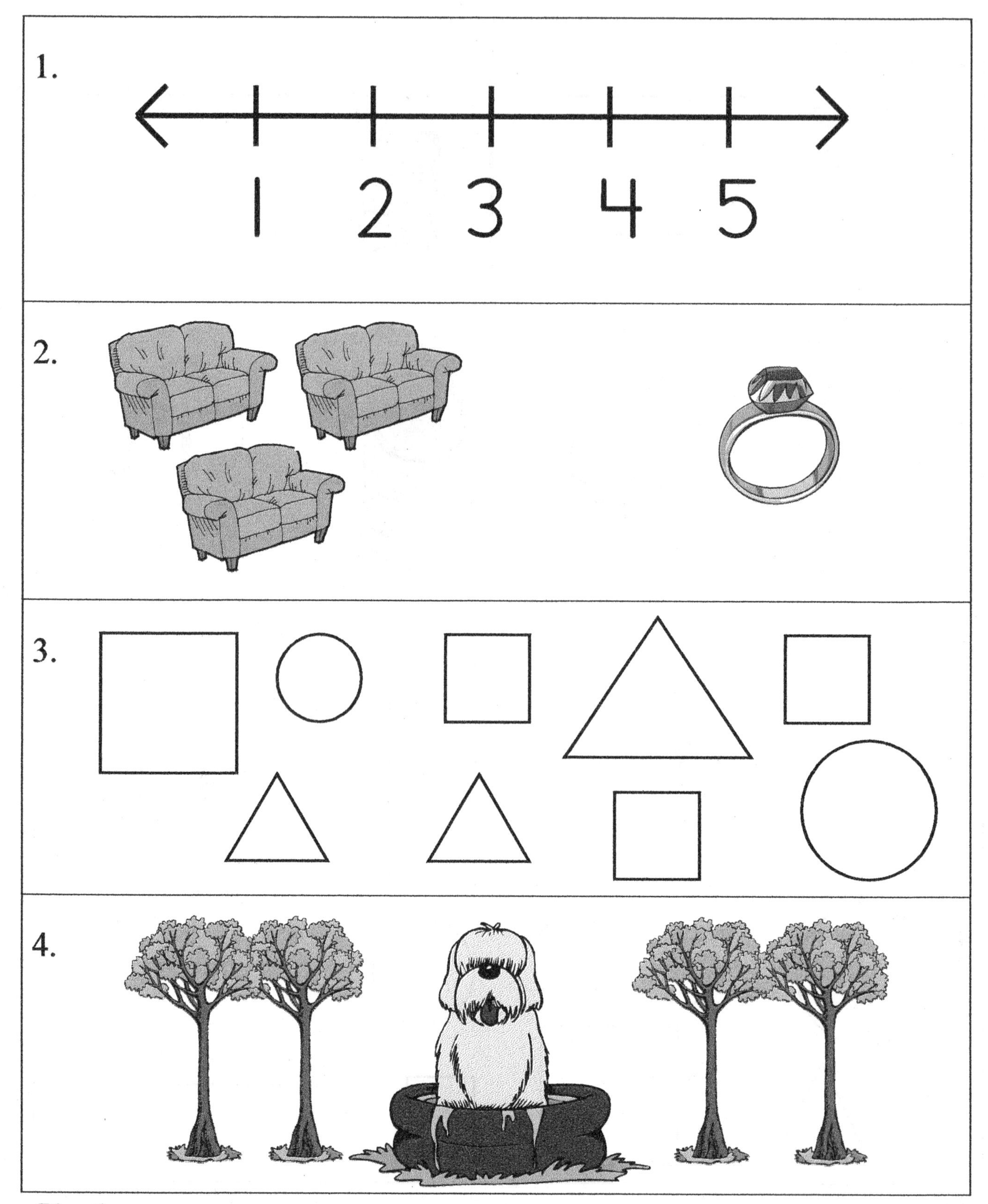

Directions:

1) Circle the number before 4.
2) Count the number in each group. Circle the group that has fewer.
3) Color the squares blue.
4) Cross out what is inside the pool.

Lesson #58

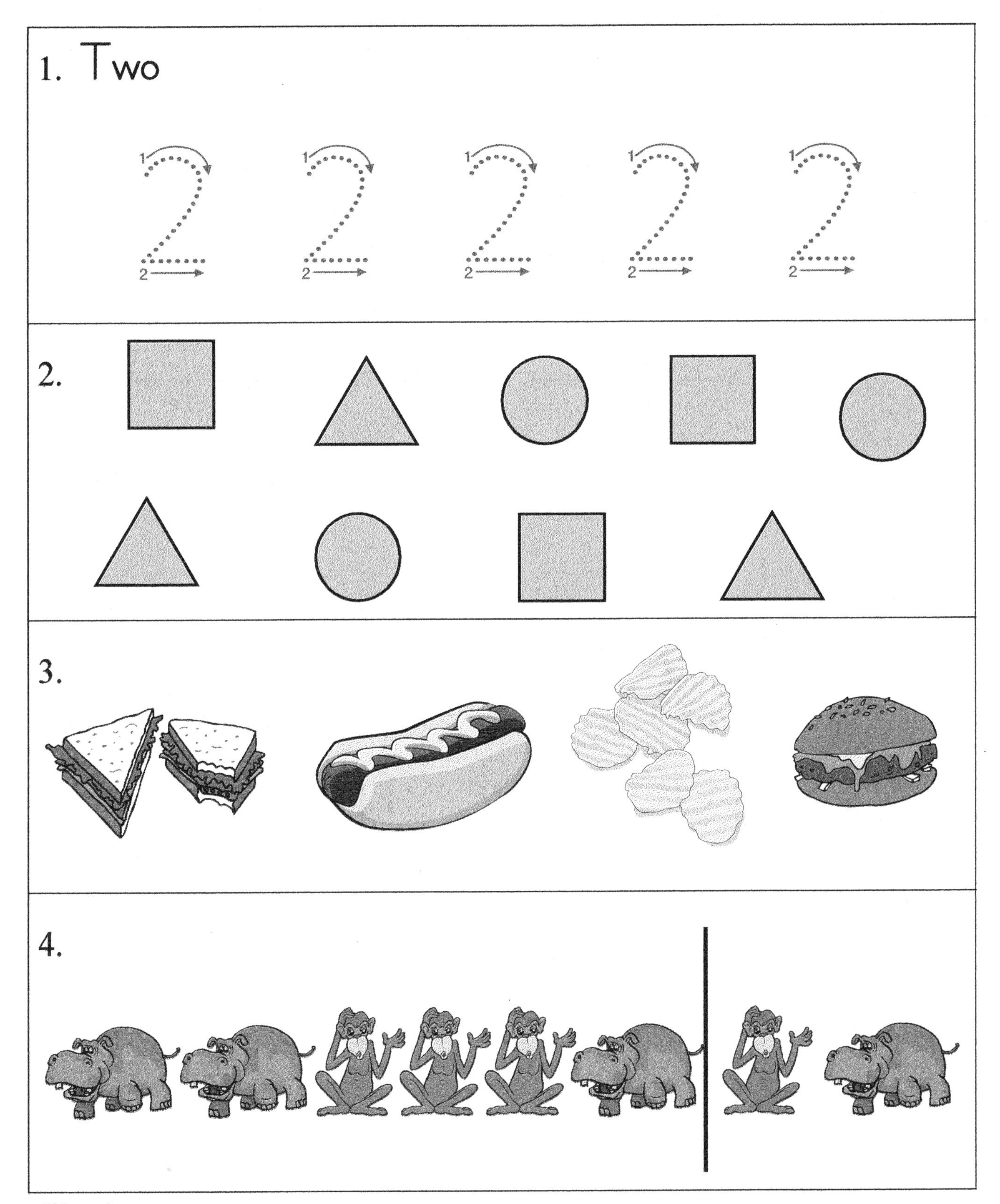

Directions:

1) Trace the number.
2) Cross out the triangles.
3) Cross out the food that doesn't belong.
4) Circle the one that should come next in the pattern.

Lesson #59

Directions:

1) Circle the shape that should come next.
2) Count the number of animals. Write the number.
3) Circle the number that comes after 3.
4) Count the number in each group. Circle the group with more.

Lesson #60

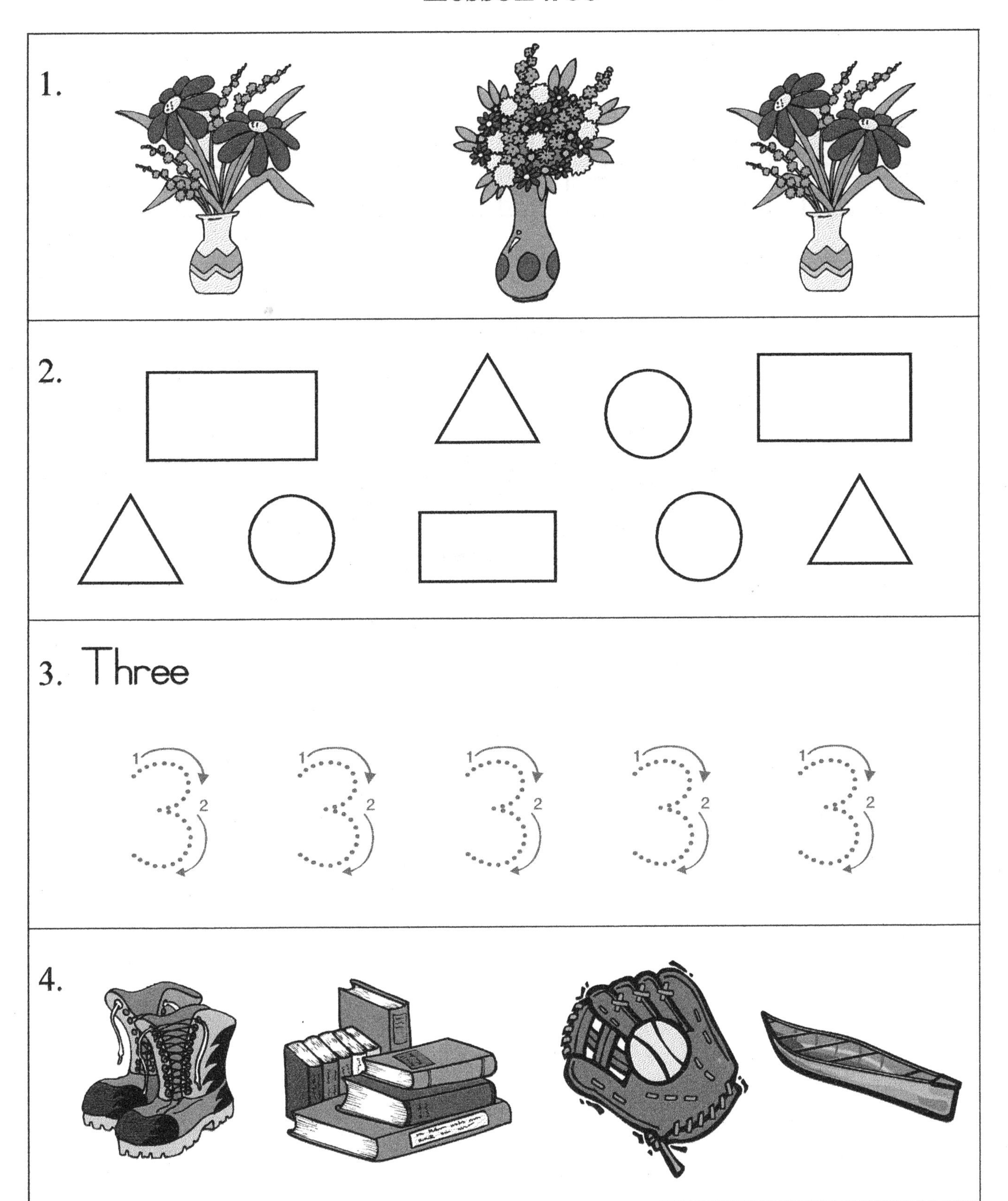

Directions:

1) Cross out the one that is different.
2) Color the rectangles orange.
3) Trace the number.
4) Circle the picture after the books.

Lesson #61

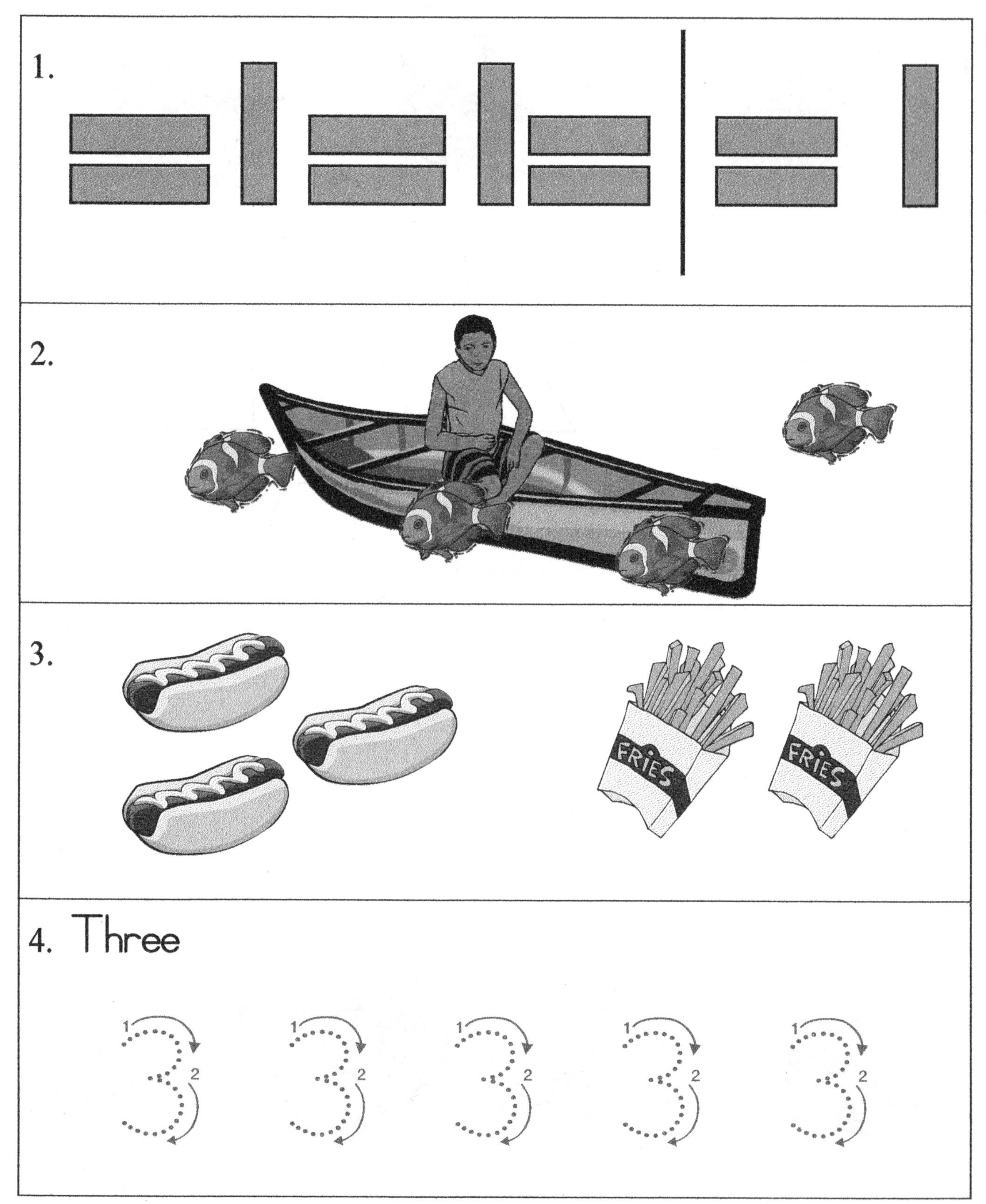

Directions:

1) Circle the shape that should come next.
2) Cross out the objects outside the boat.
3) Count the number in each group. Circle the group with fewer.
4) Trace the number.

Lesson #62

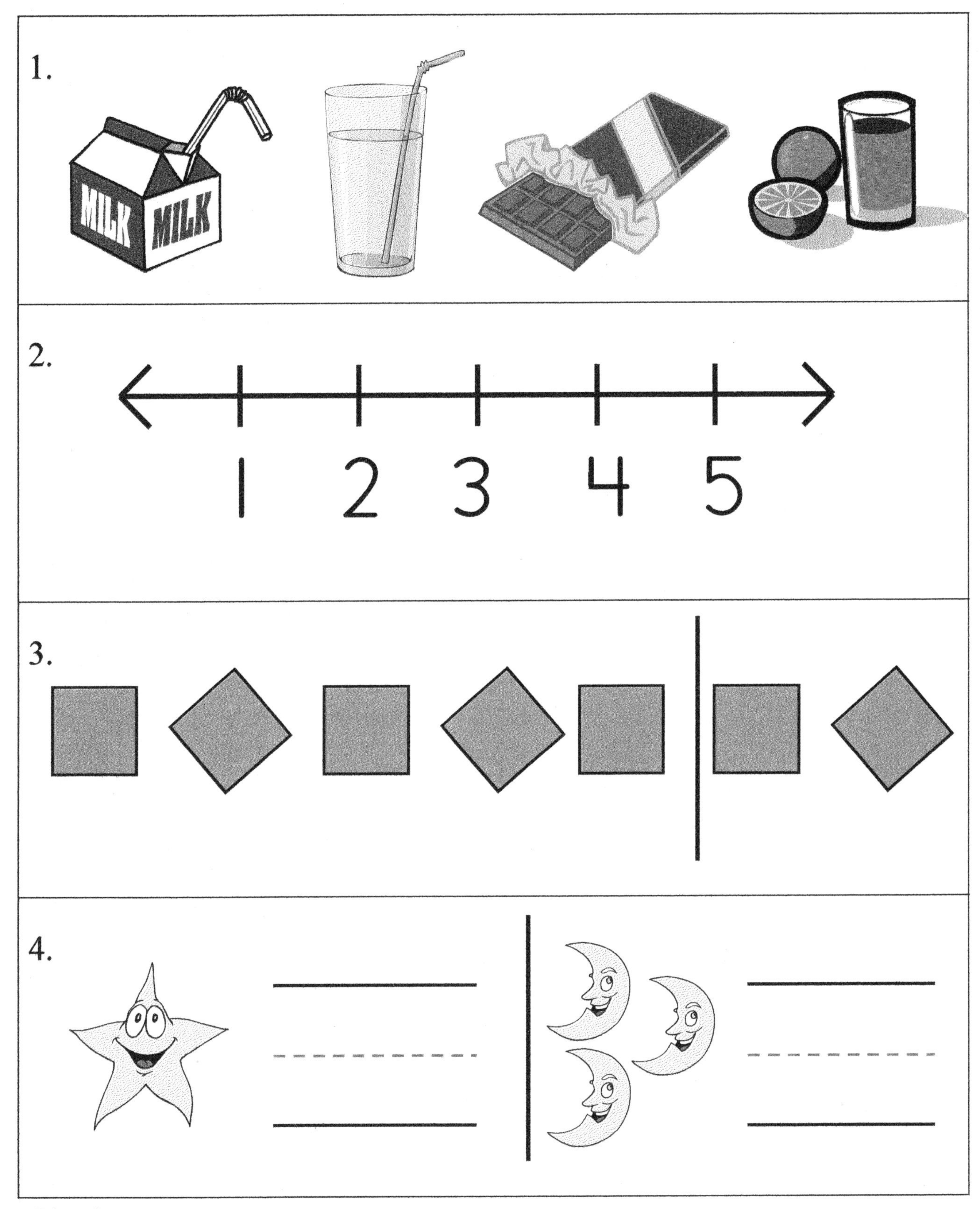

Directions:

1) Cross out the one that doesn't belong.
2) Circle the number that comes after 1.
3) Circle the shape that should come next in the pattern.
4) Count the number in each group. Write the number.

Lesson #63

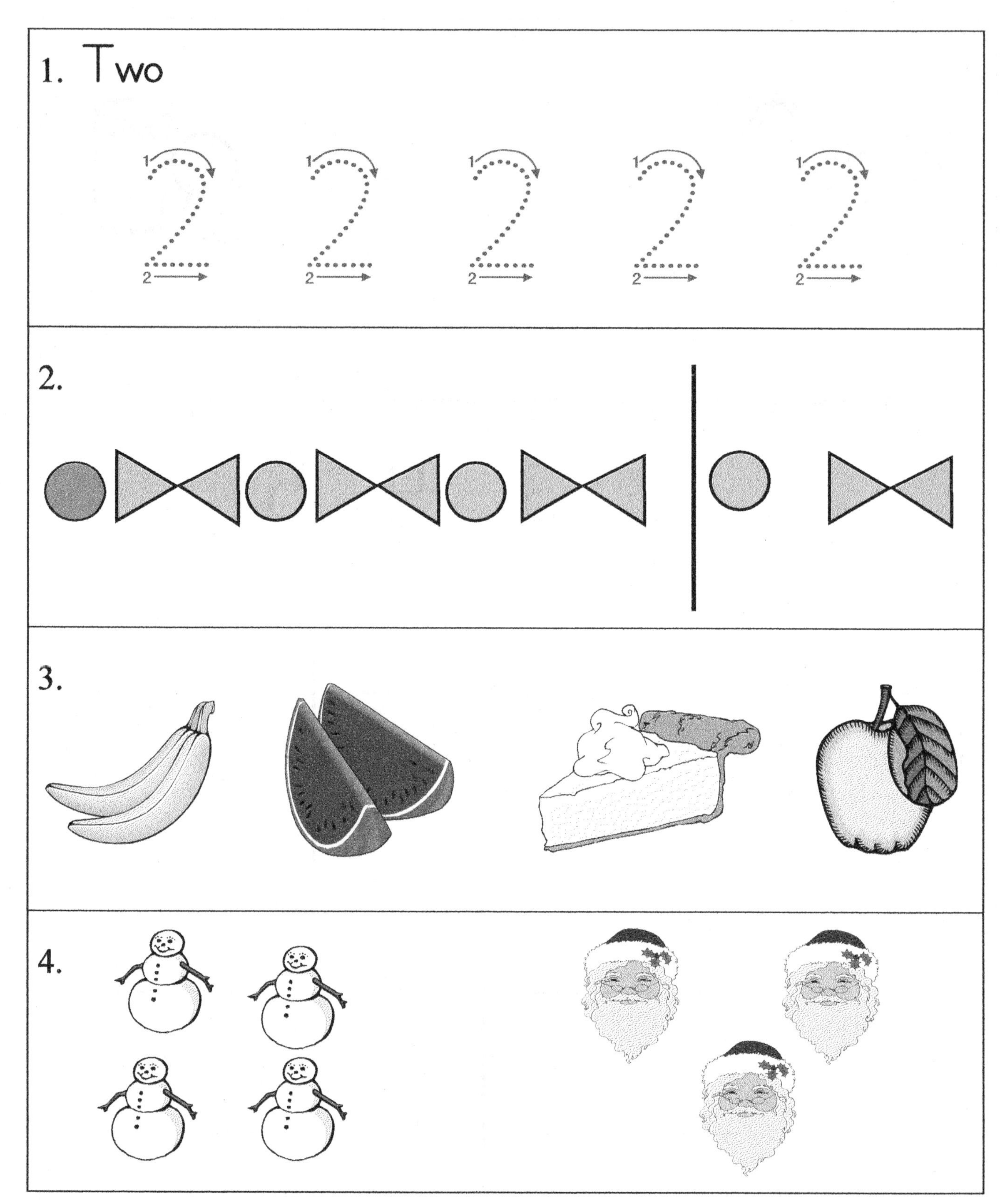

Directions:

1) Trace the number.
2) Circle the shape that should come next.
3) Cross out the food that doesn't belong.
4) Count the number in each group. Circle the group that has more.

Lesson #64

Directions:

1) Count the number of scooters. Write the number three times.
2) Color the squares green.
3) Circle the number that comes after 4.
4) Circle the animal that should come next in the pattern.

Lesson #65

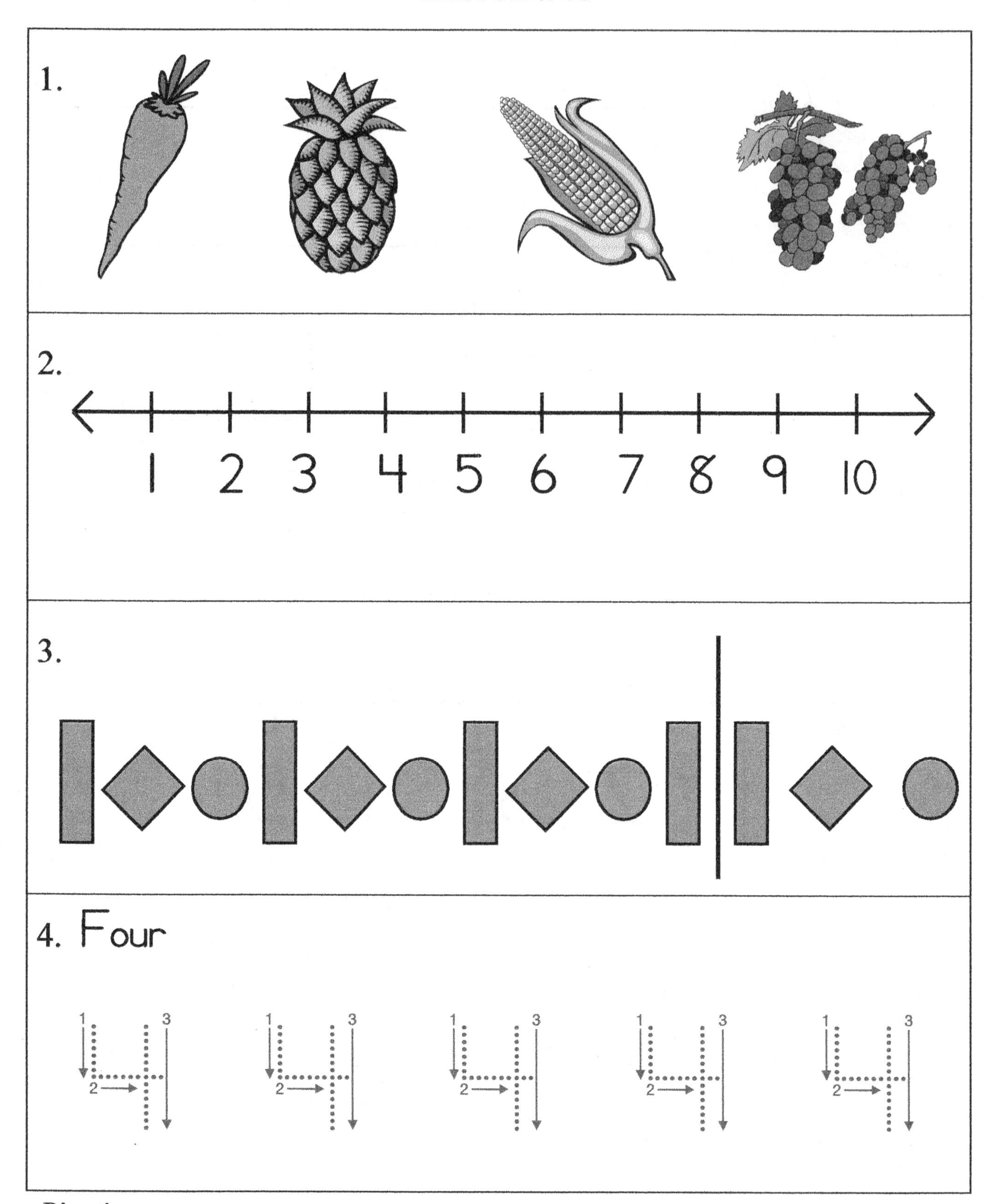

Directions:

1) Cross out the picture after the carrot.
2) Circle the number that comes before 7.
3) Circle the shape that should come next in the pattern.
4) Trace the number.

Lesson #66

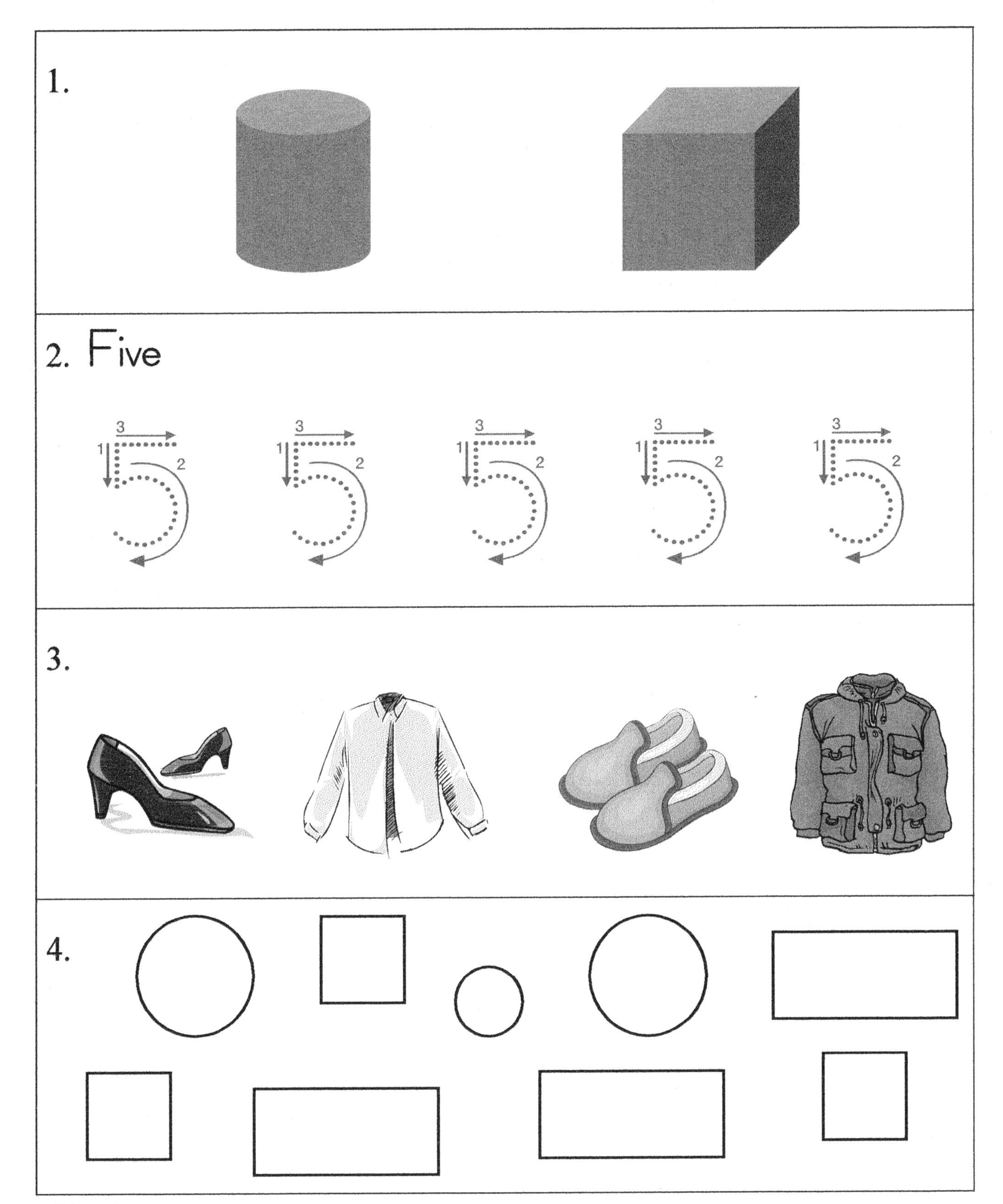

Directions:

1) Circle the shape that can roll.
2) Trace the number.
3) Cross out the things a person can wear on their feet.
4) Circle the rectangles blue.

Lesson #67

1.

2.

3.

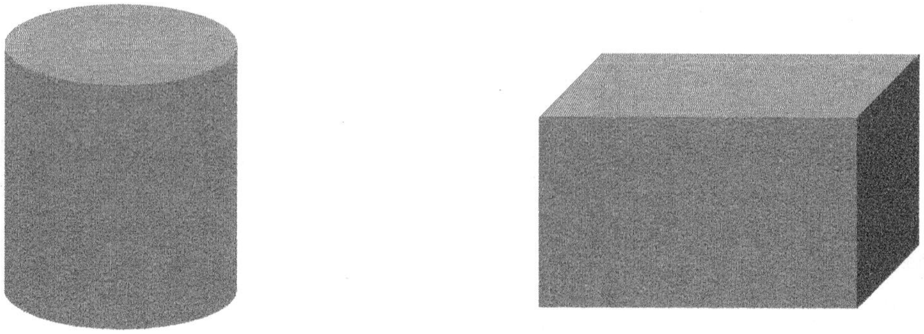

4.

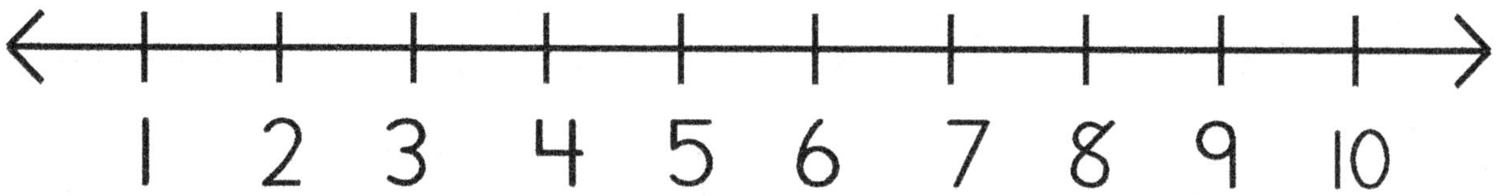

1 2 3 4 5 6 7 8 9 10

Directions:

1) Count the number of shoes. Write the number three times.
2) Circle the one that should come next.
3) Cross out the shape with corners.
4) Circle the number that comes after **8**.

Lesson #68

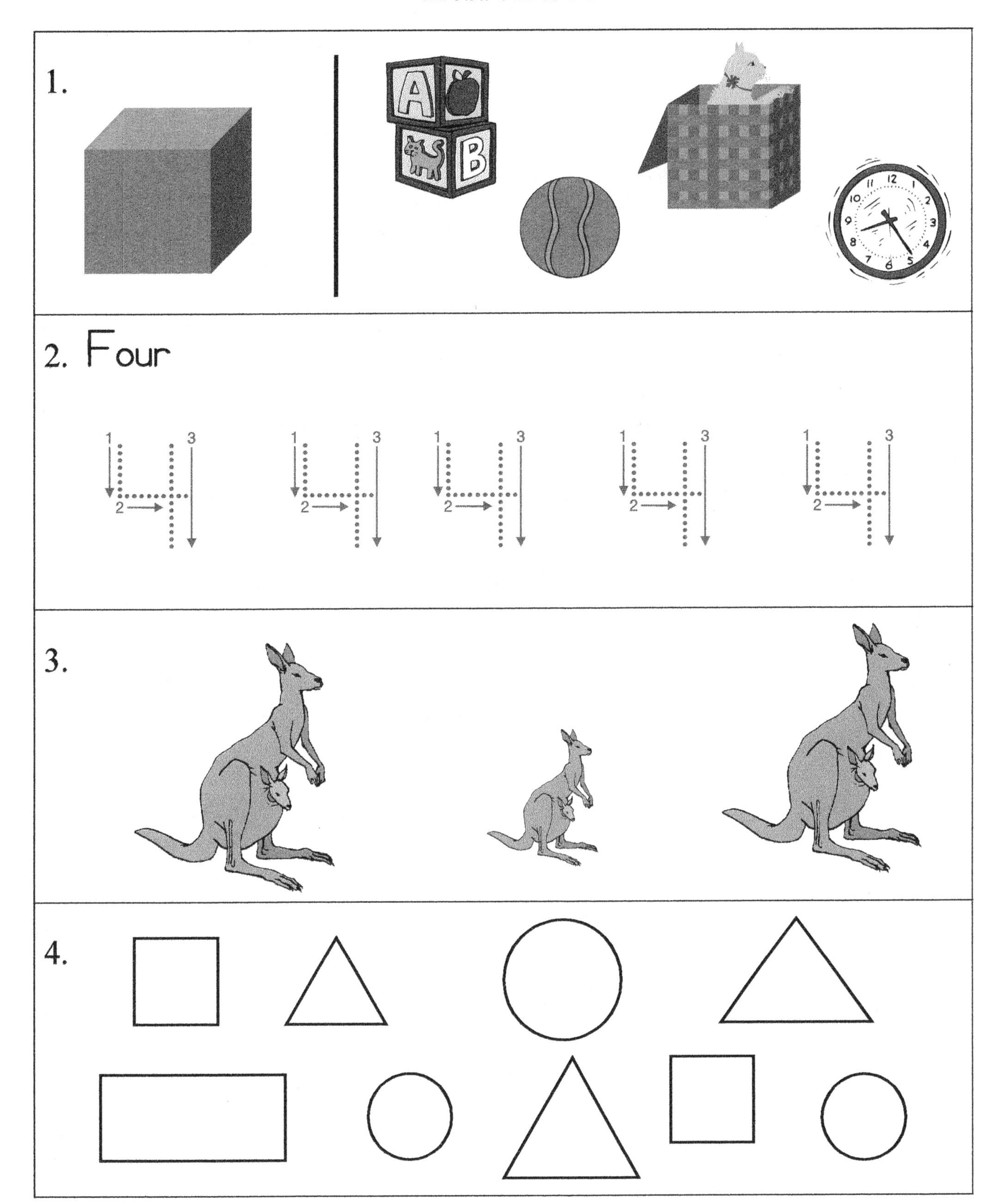

Directions:

1) Look at the shape before the line. Circle the objects that are like that shape.
2) Trace the number.
3) Cross out the one that is different.
4) Color the triangles red.

Lesson #69

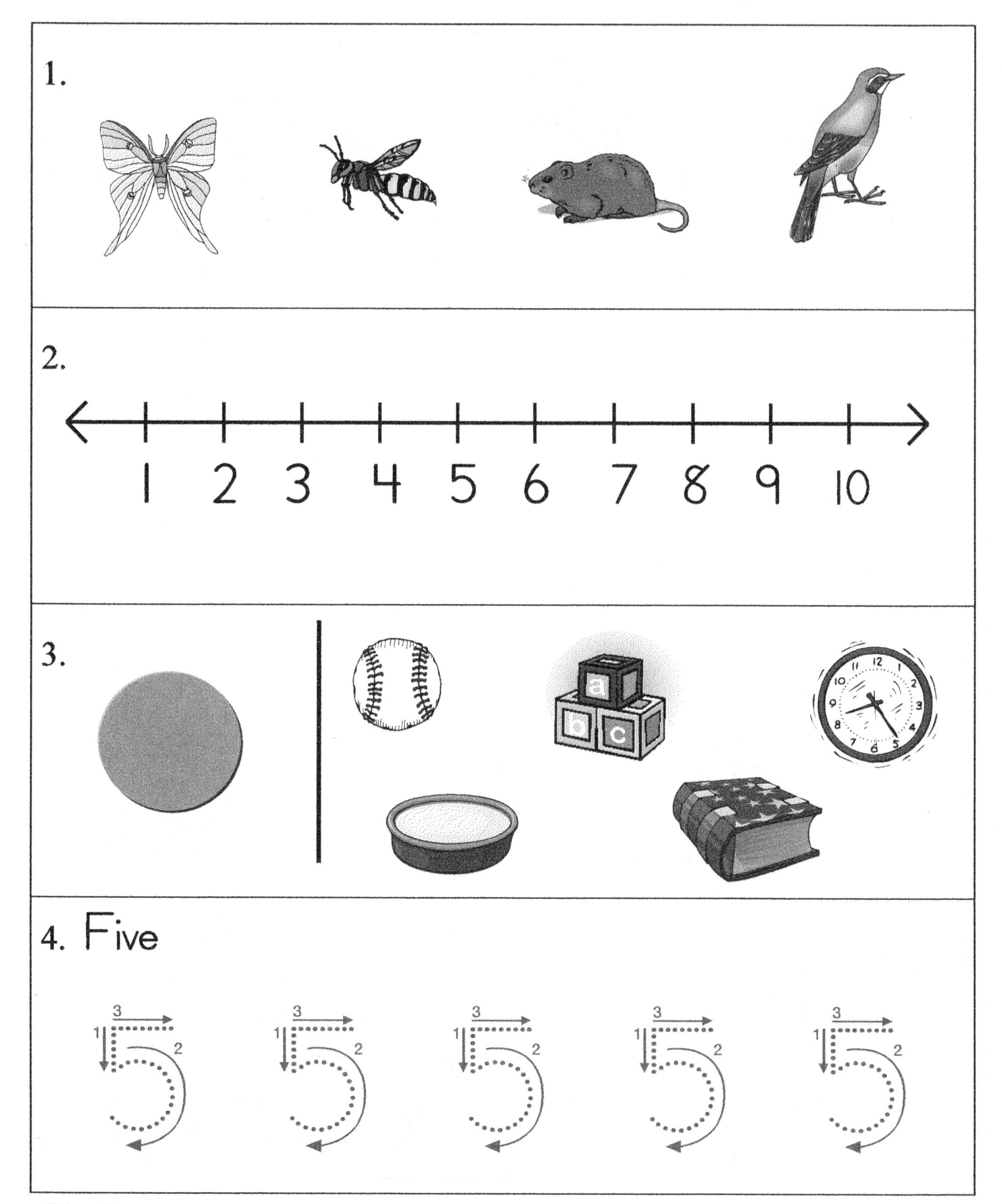

Directions:

1) Cross out the one that doesn't belong.
2) Circle the number that comes before **8**.
3) Look at the shape before the line. Circle the objects that are like that shape.
4) Trace the number.

Lesson #70

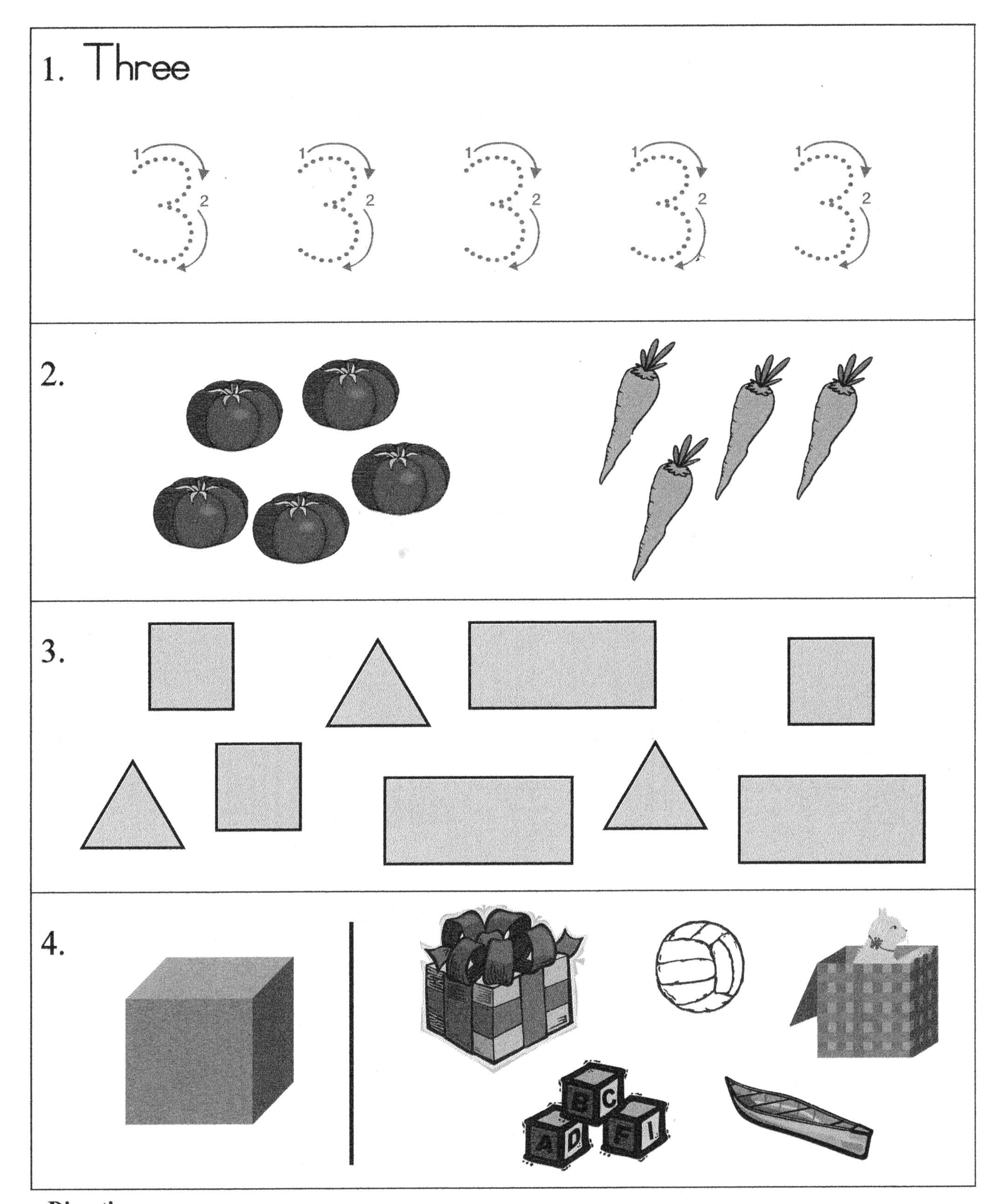

Directions:

1) Trace the number.
2) Count the number in each group. Circle the group that has fewer.
3) Color the squares green.
4) Cross out the objects that are like the shape before the line.

Lesson #71

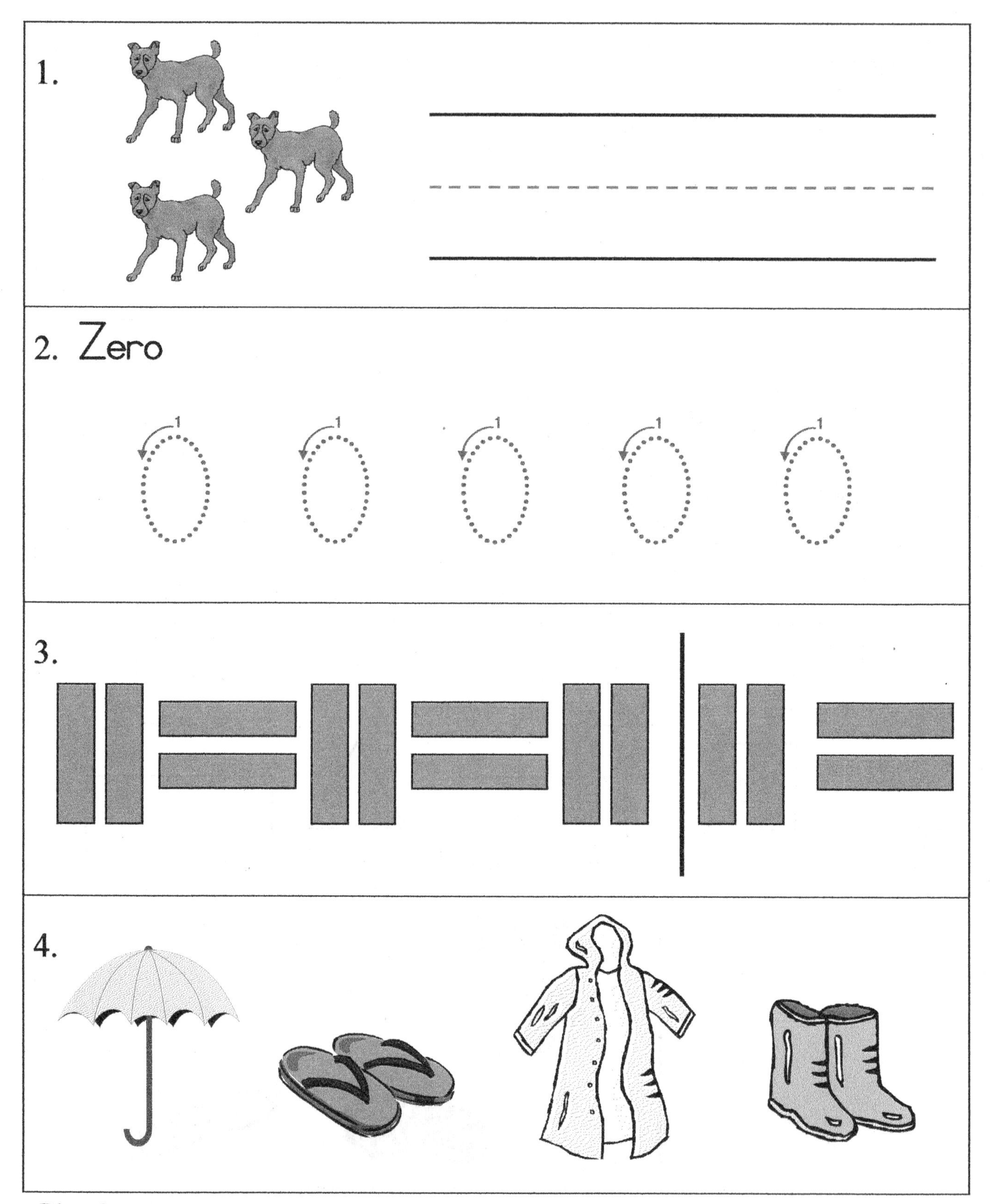

Directions:

1) Count the number of dogs. Write the number.
2) Trace the number.
3) Circle the shape that should come next.
4) Cross out the picture that doesn't belong.

Lesson #72

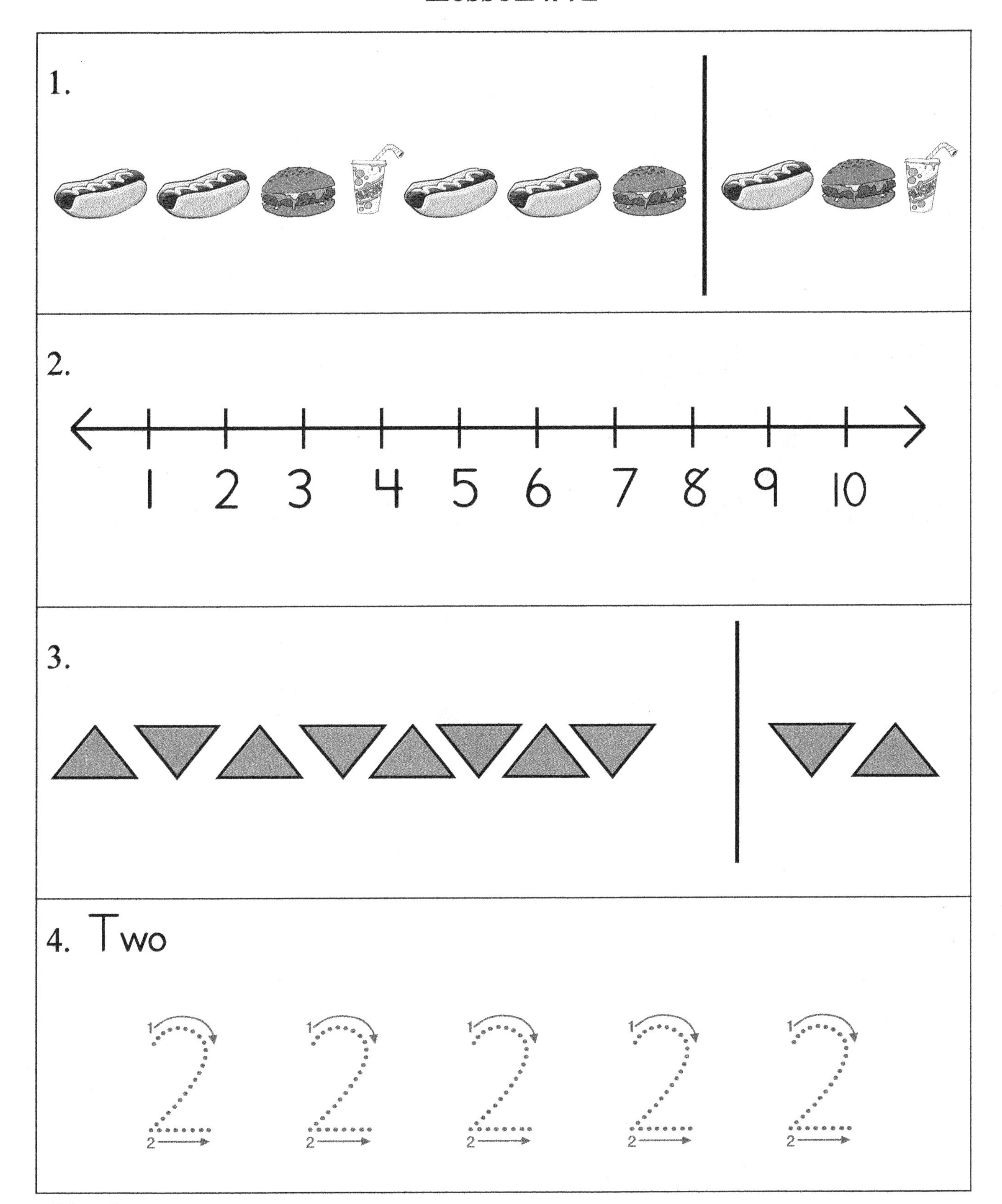

Directions:

1) Cross out the one that is likely to come next.
2) Circle the number that comes after 5.
3) Circle the shape that comes next.
4) Trace the number.

Lesson #73

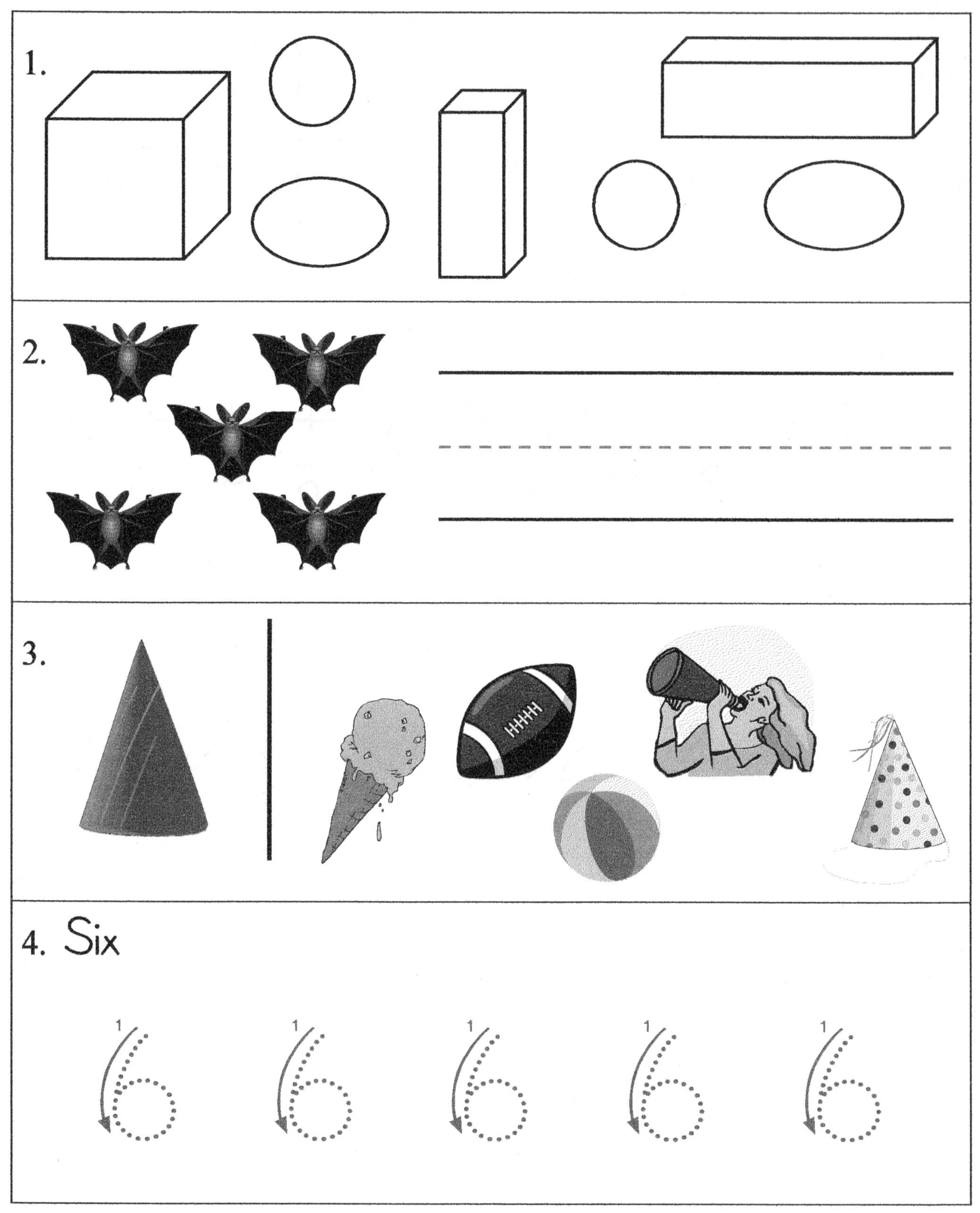

Directions:

1) Color the shapes with corners using orange.
2) Count the number of bats. Write the number.
3) Circle any picture that includes a shape like the one before the line.
4) Trace the number.

Lesson #74

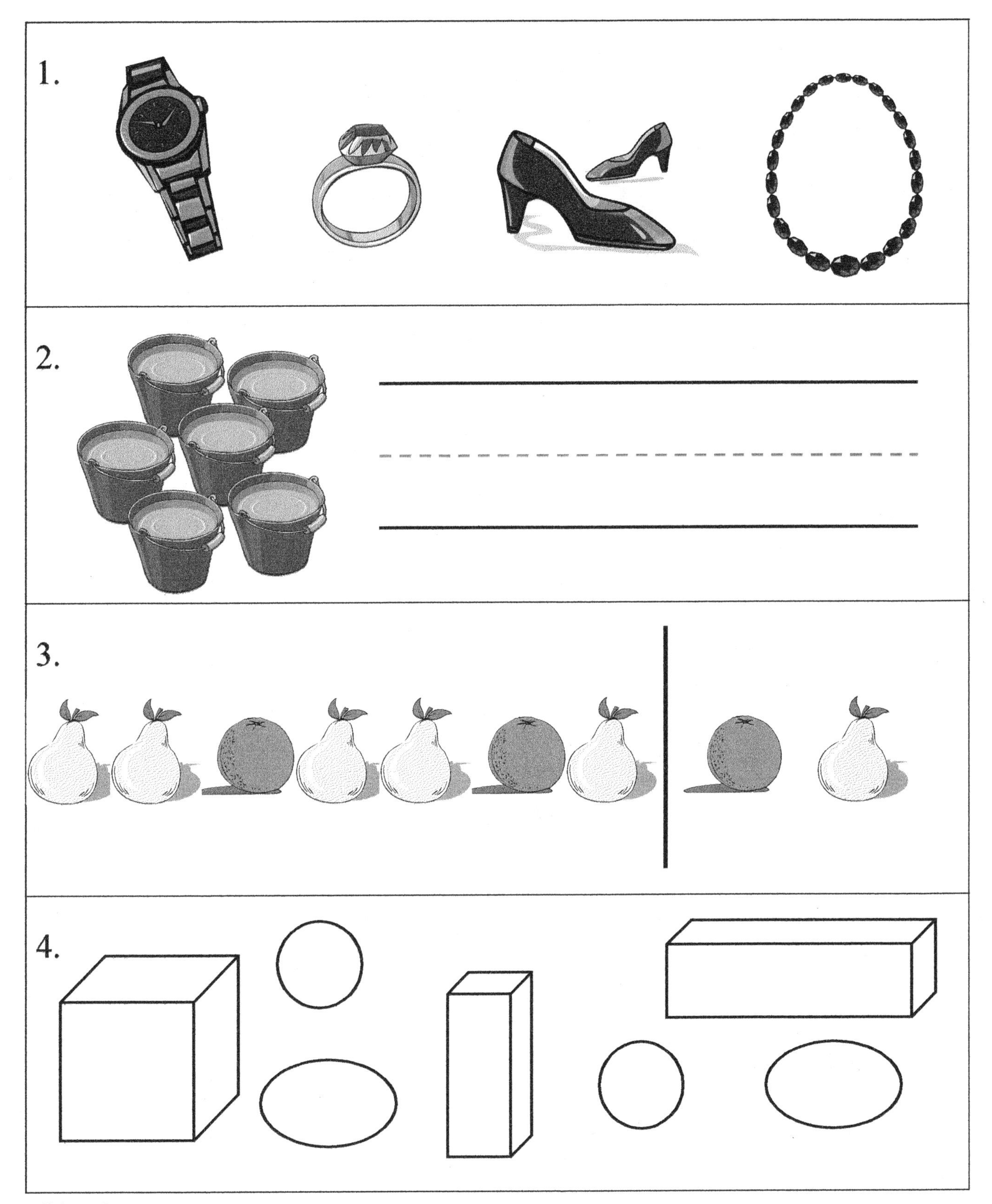

Directions:

1) Cross out the picture that doesn't belong.
2) Count the number of buckets. Write the number.
3) Circle the one that should come next.
4) Find the shapes without corners. Color them blue.

Lesson #75

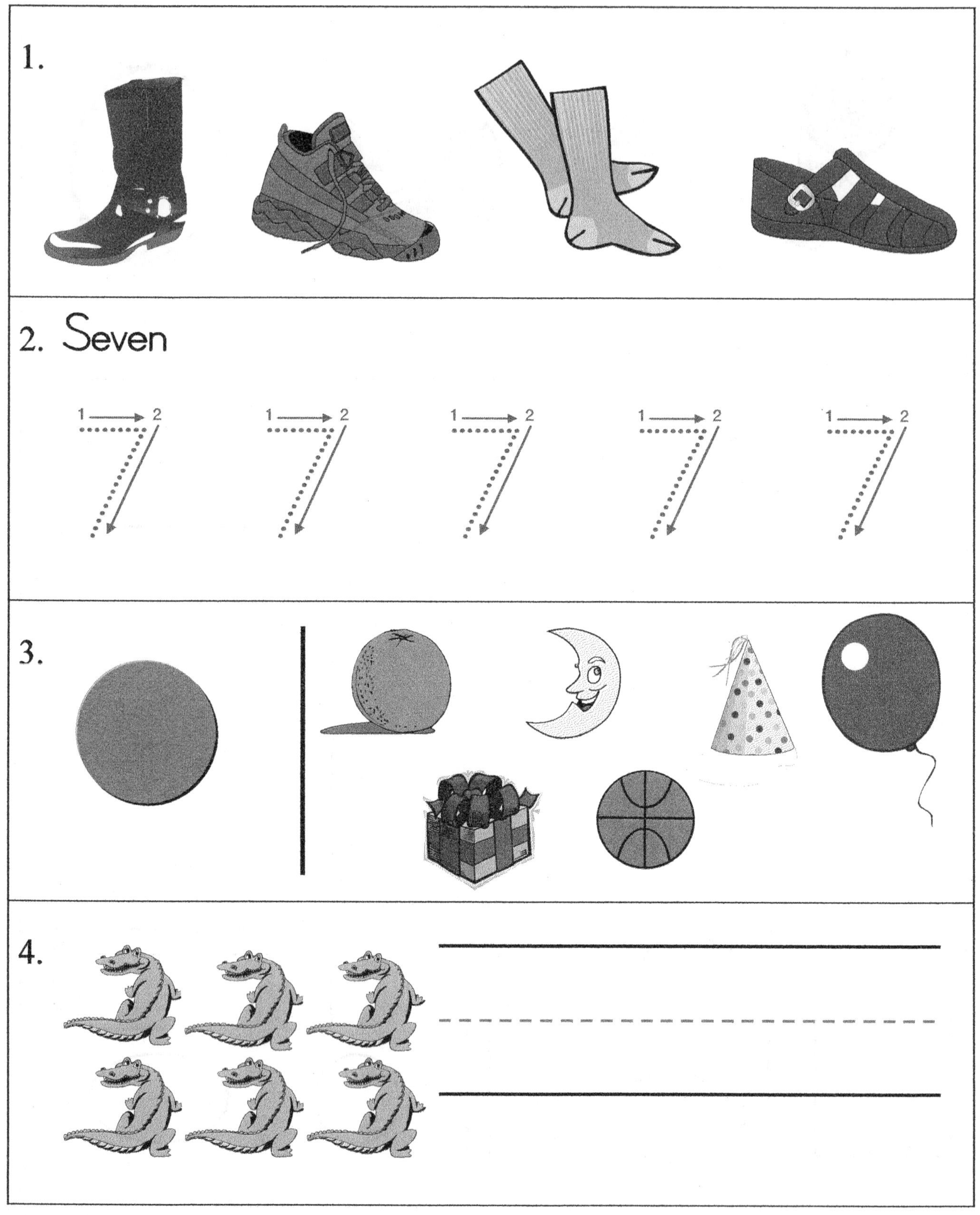

Directions:

1) Cross out the picture that doesn't belong.
2) Trace the number.
3) Circle the objects that are like the solid shape.
4) Count the number of crocodiles. Write the number.

Lesson #76

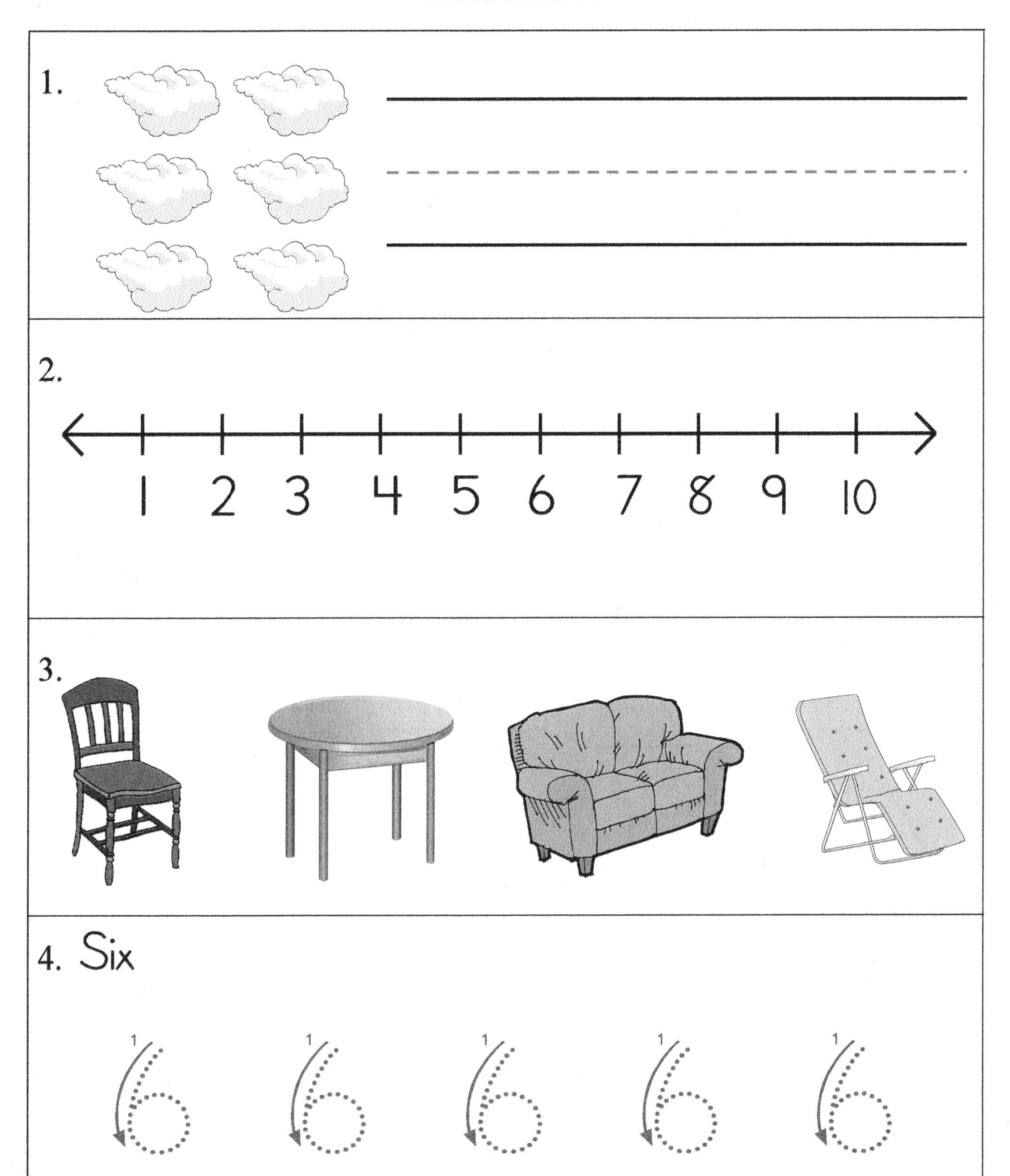

Directions:

1) Count the number of clouds. Write the number.
2) Circle the number after 9.
3) Cross out the picture that doesn't belong.
4) Trace the number.

Lesson #77

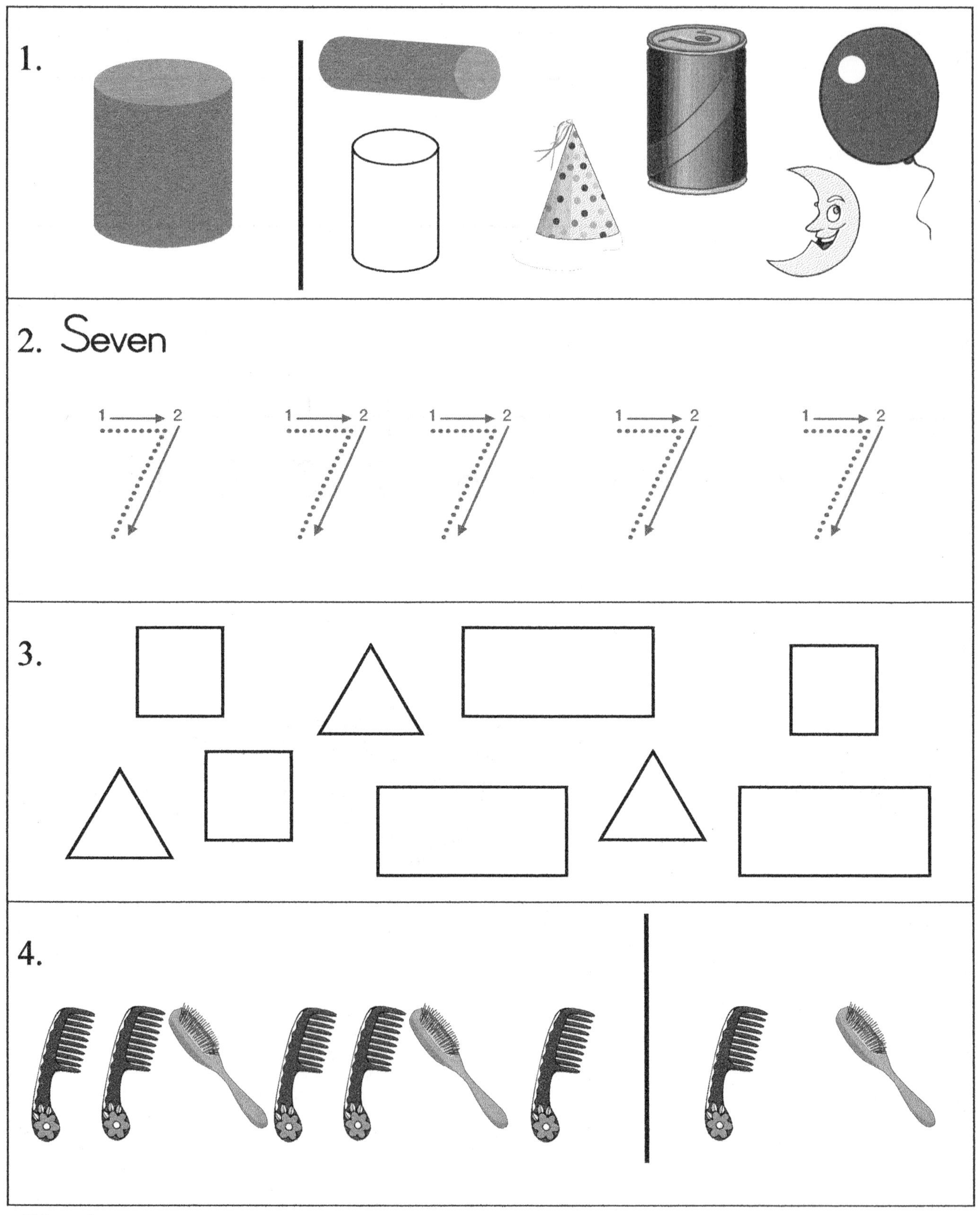

Directions:

1) Cross out the objects that are like the solid shape.
2) Trace the number.
3) Color the rectangles red.
4) Circle the object that should come next.

Lesson #78

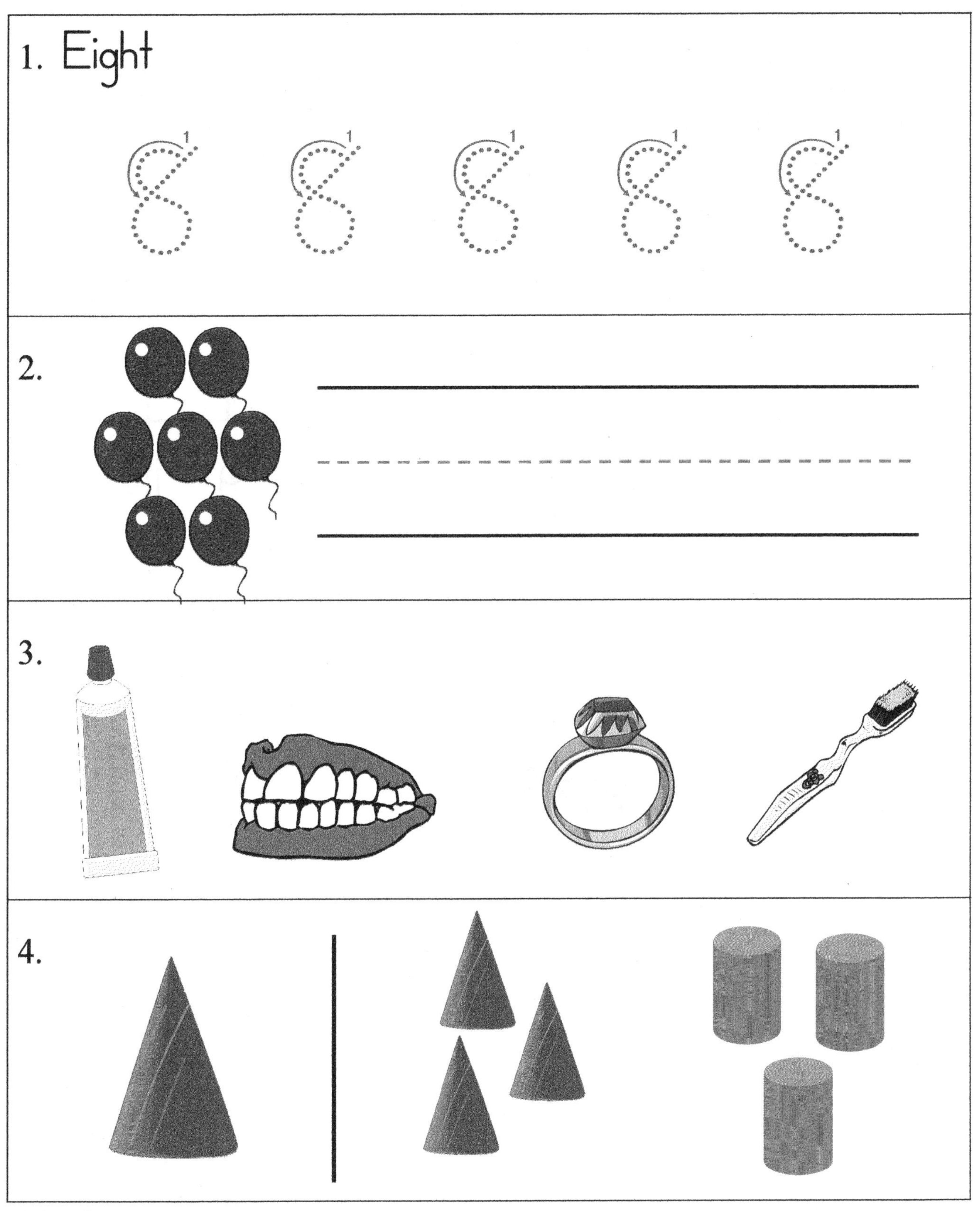

Directions:

1) Trace the number.
2) Count the balloons. Write the number.
3) Cross out the one that doesn't belong.
4) Circle the group where the shape belongs.

Lesson #79

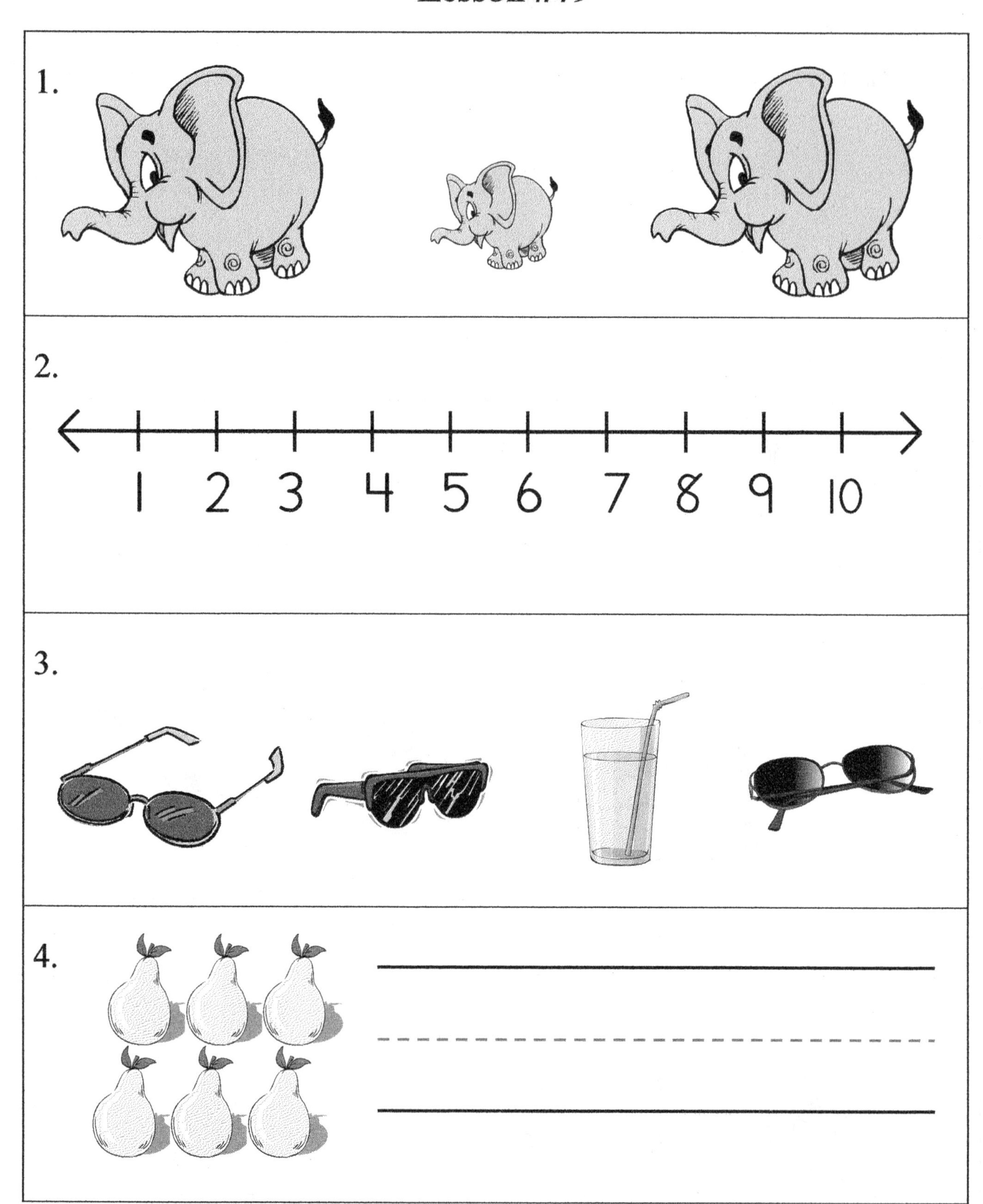

Directions:

1) Cross out the one that doesn't belong.
2) Circle the number before 10.
3) Cross out the picture that doesn't belong.
4) Count the number of pears. Write the number.

Lesson #80

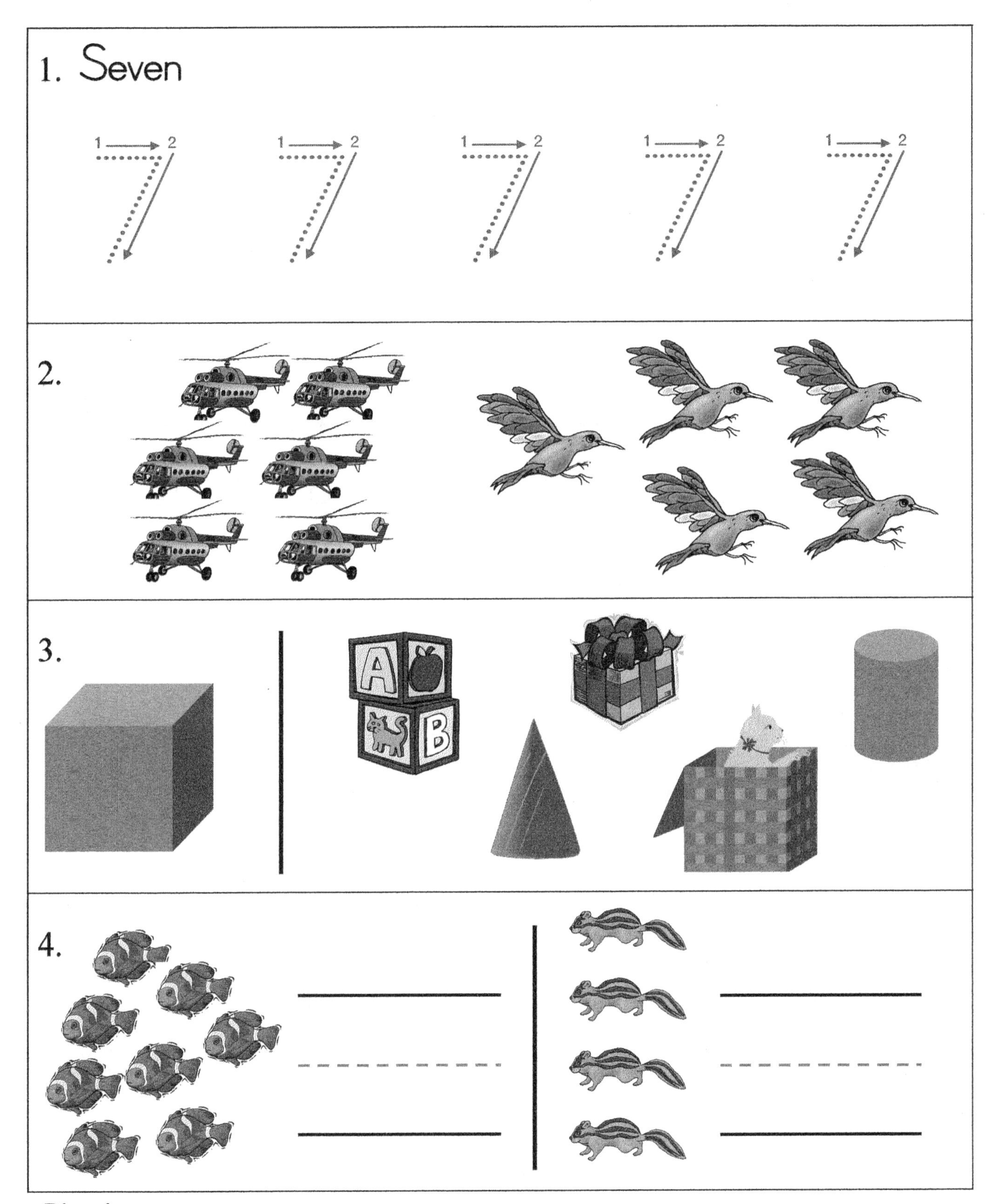

Directions:

1) Trace the number.
2) Circle the group that has fewer.
3) Circle the objects that are like the solid shape.
4) Count the number in each group. Write the number.

Lesson #81

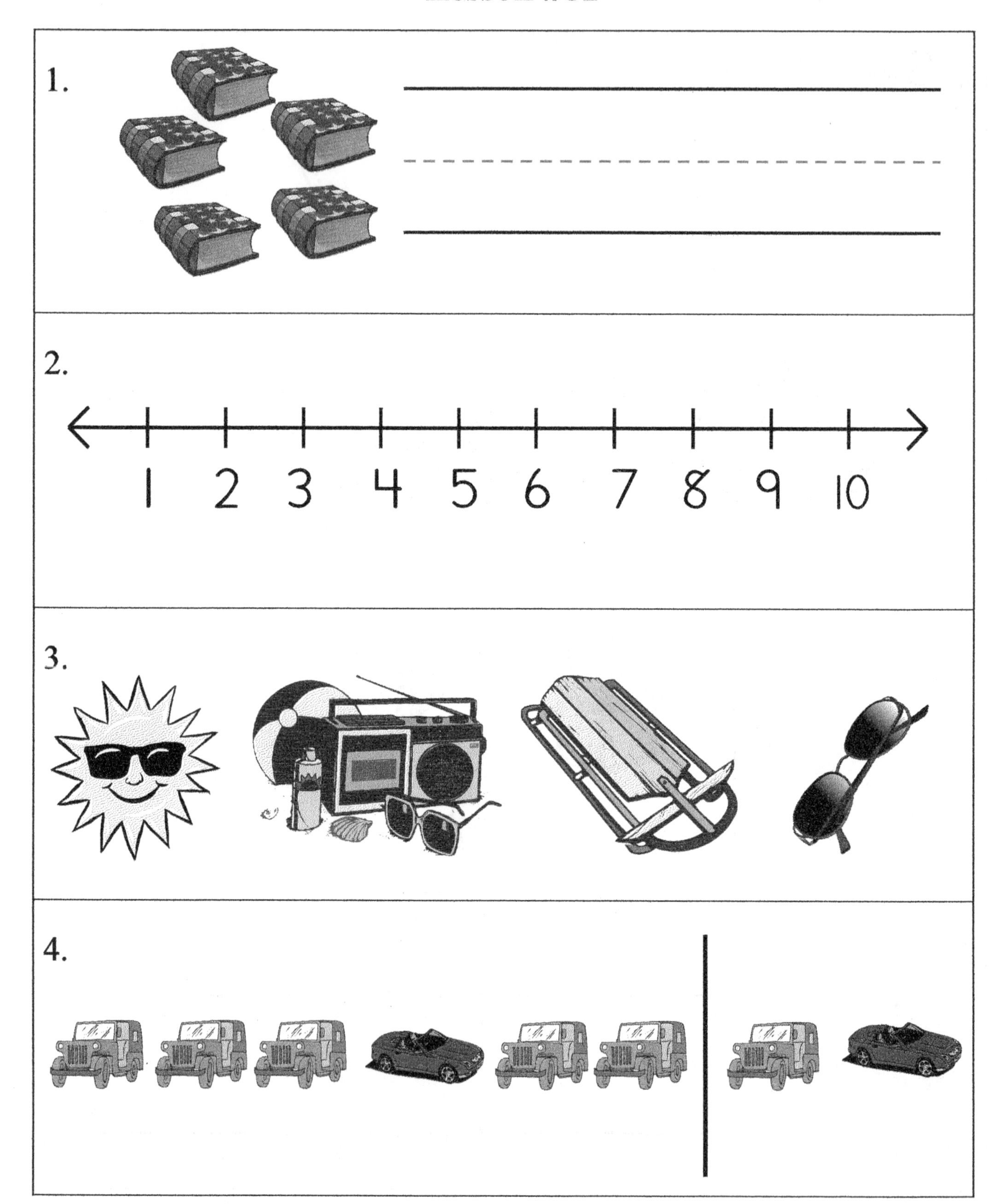

Directions:

1) Count the number of books. Write the number.
2) Circle the number before 9.
3) Cross out the picture that doesn't belong.
4) Circle the one that should come next.

Lesson #82

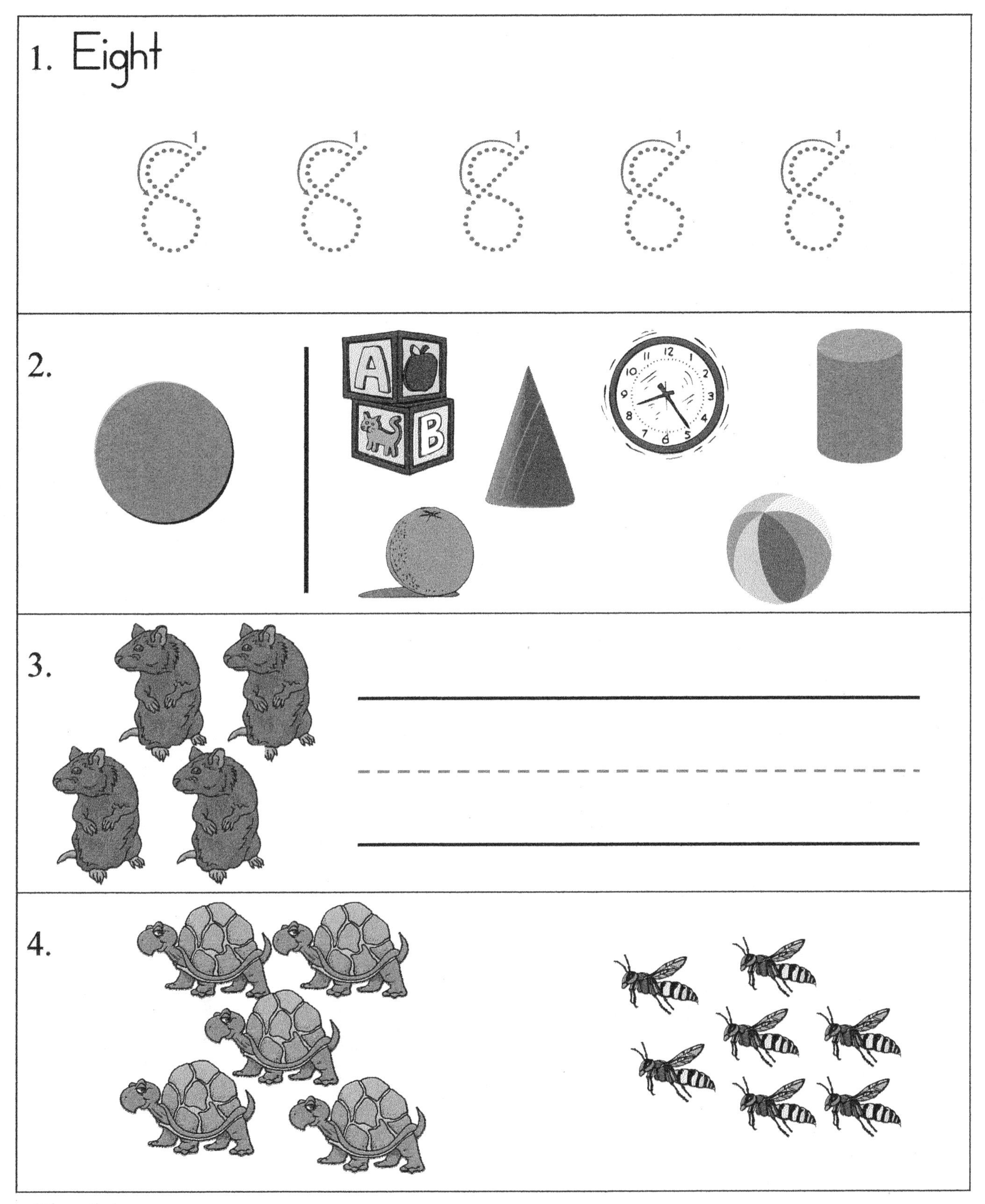

Directions:

1) Trace the number.
2) Circle the objects that are like the solid shape.
3) Count the hamsters. Write the number.
4) Count the number in each group. Circle the group that has more.

Lesson #83

Directions:

1) Count the tigers. Write the number.
2) Trace the number.
3) Cross out the picture that doesn't belong.
4) Fill in the missing number.

Lesson #84

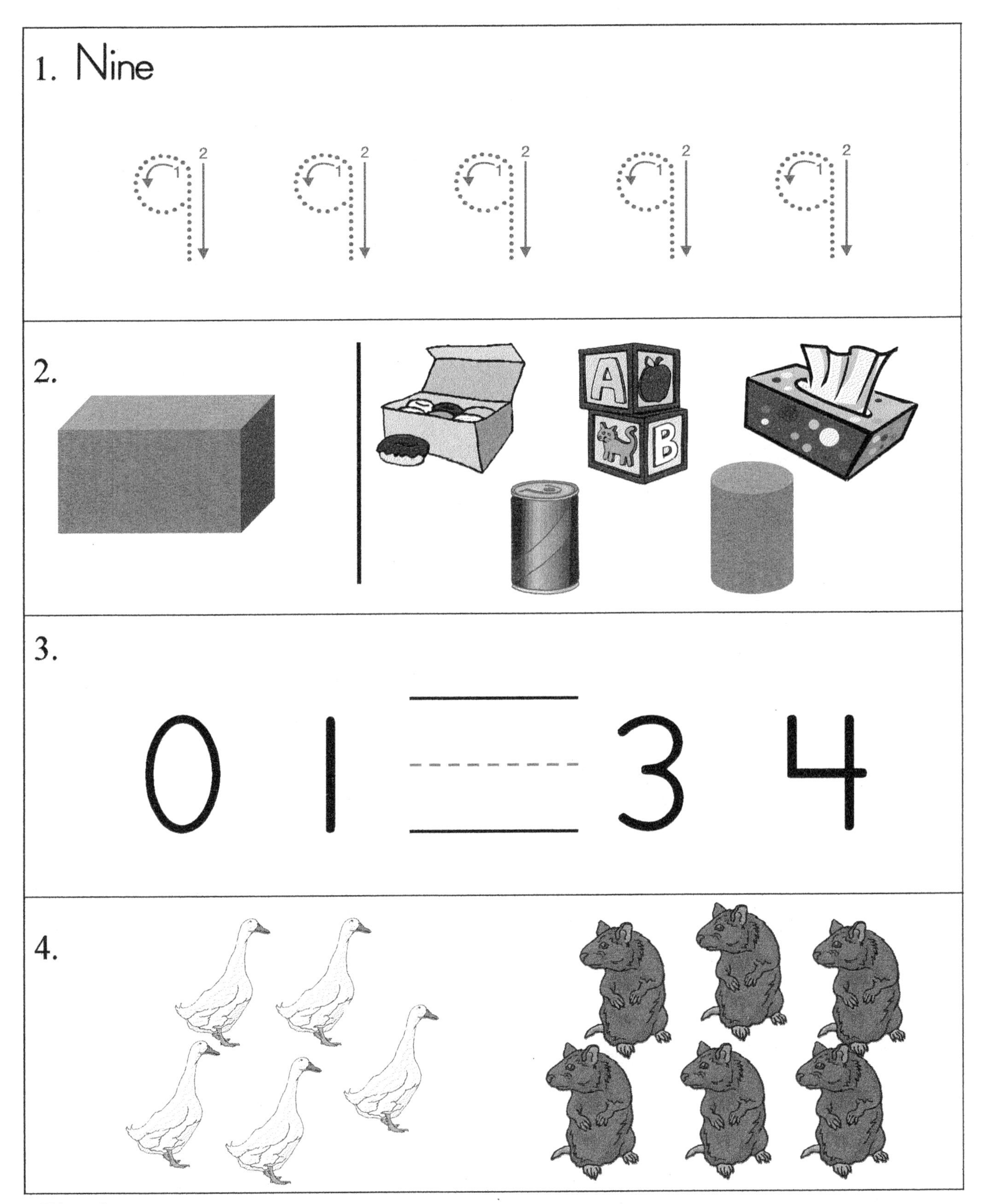

Directions:

1) Trace the number.
2) Cross out the objects that are like the solid shape.
3) Write the missing number.
4) Count the number in each group. Circle the group that has fewer.

Lesson #85

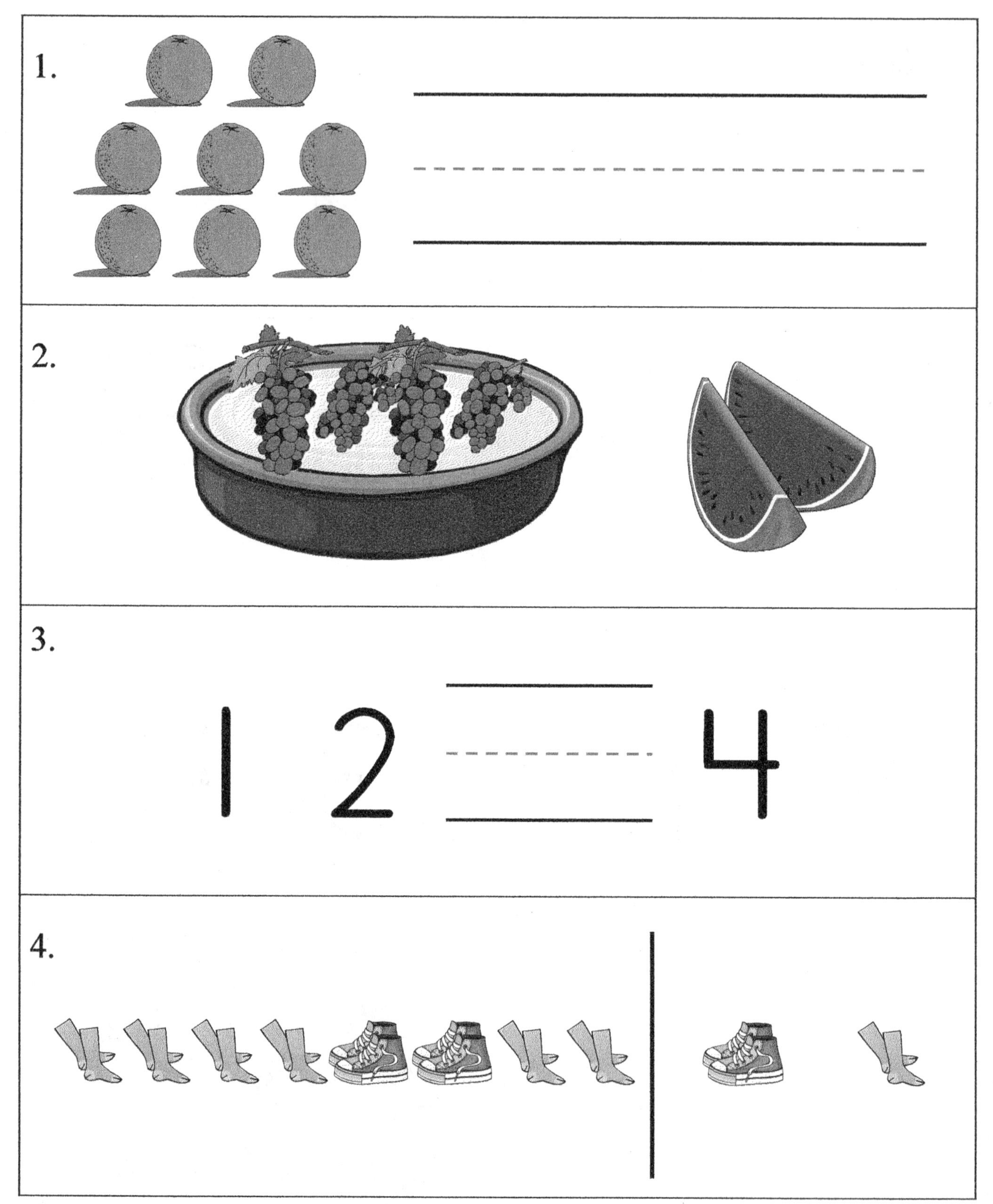

Directions:

1) Count the number of oranges. Write the number.
2) Circle the fruit that is outside the bowl.
3) Write the missing number.
4) Circle the one that is likely to come next.

Lesson #86

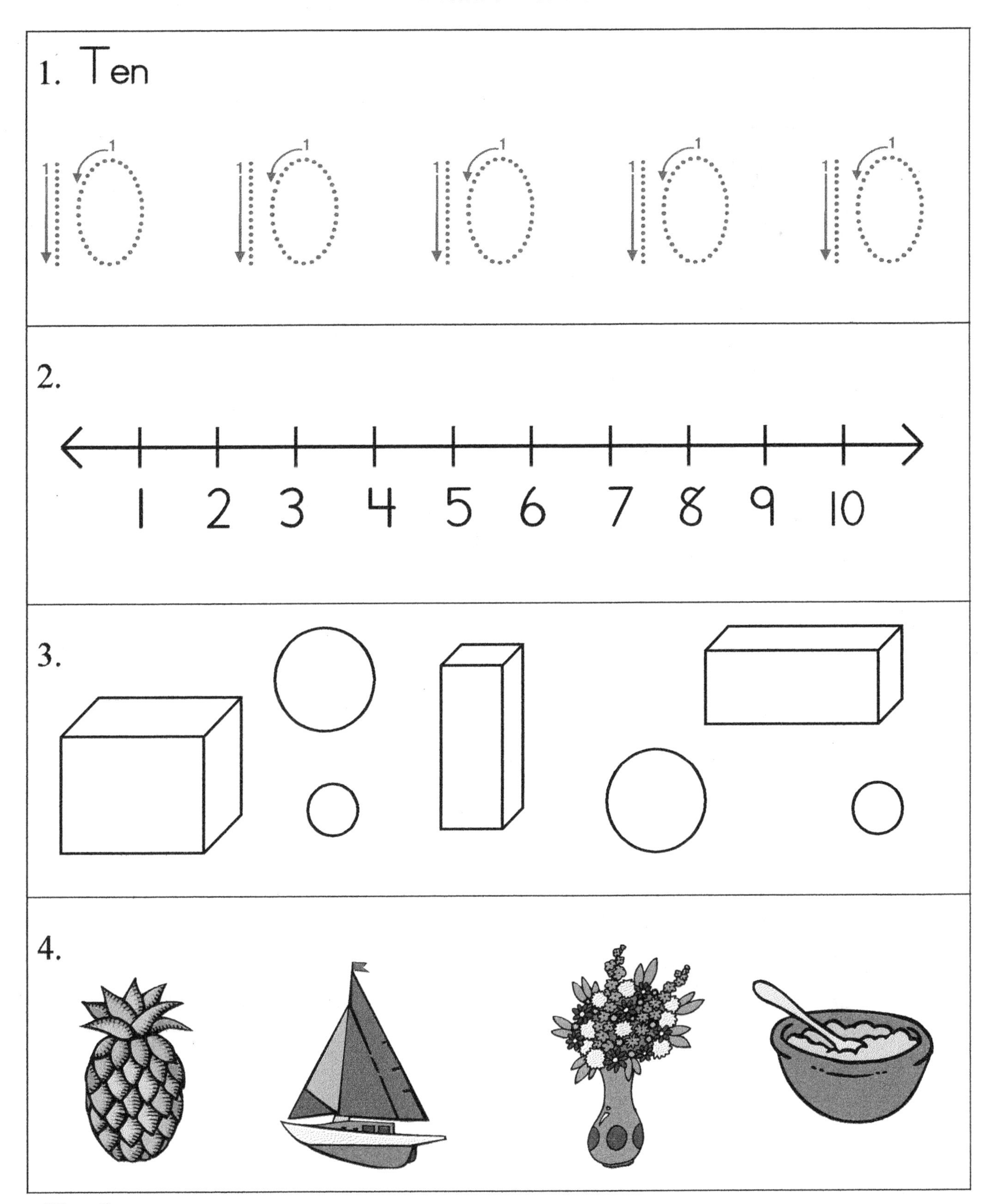

Directions:

1) Trace the number.
2) Circle the number after 9.
3) Color the shapes with corners in blue.
4) Cross out the picture after the boat.

Lesson #87

1. 3 4 ___ 6

2. ___ ___

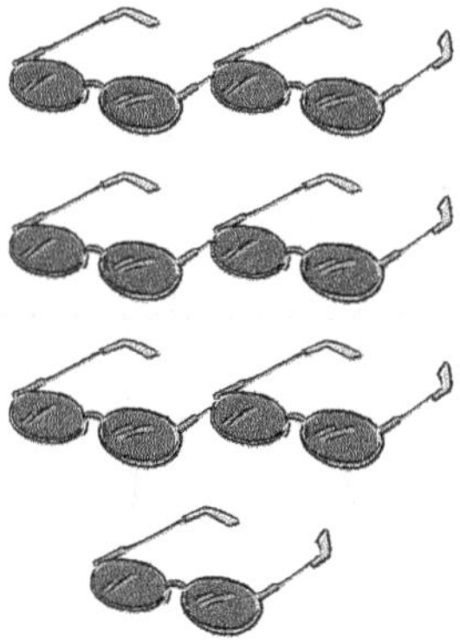

3. Ten

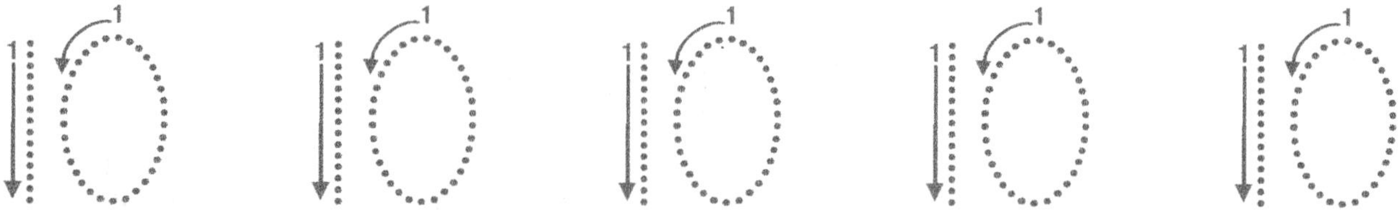

10 10 10 10 10

4.

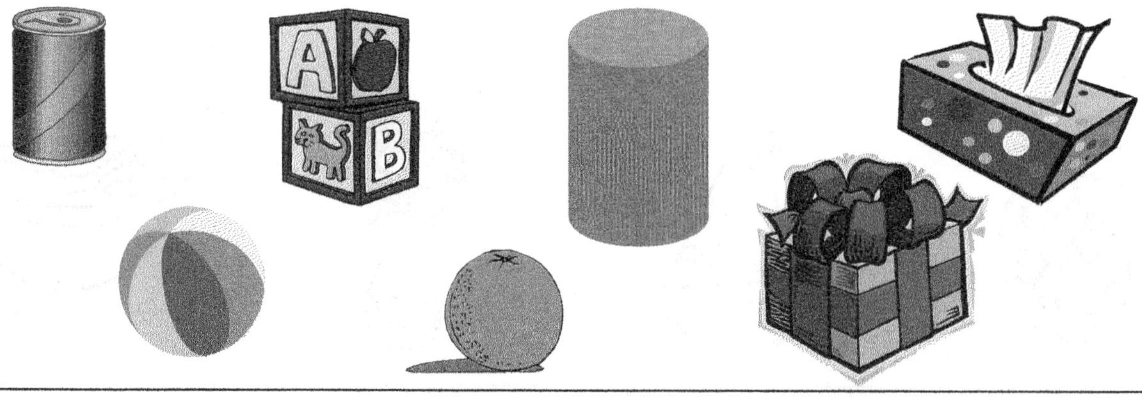

Directions:

1) Write the missing number.
2) Count the number in each group. Write the number.
3) Trace the number.
4) Cross out the objects that can roll.

Lesson #88

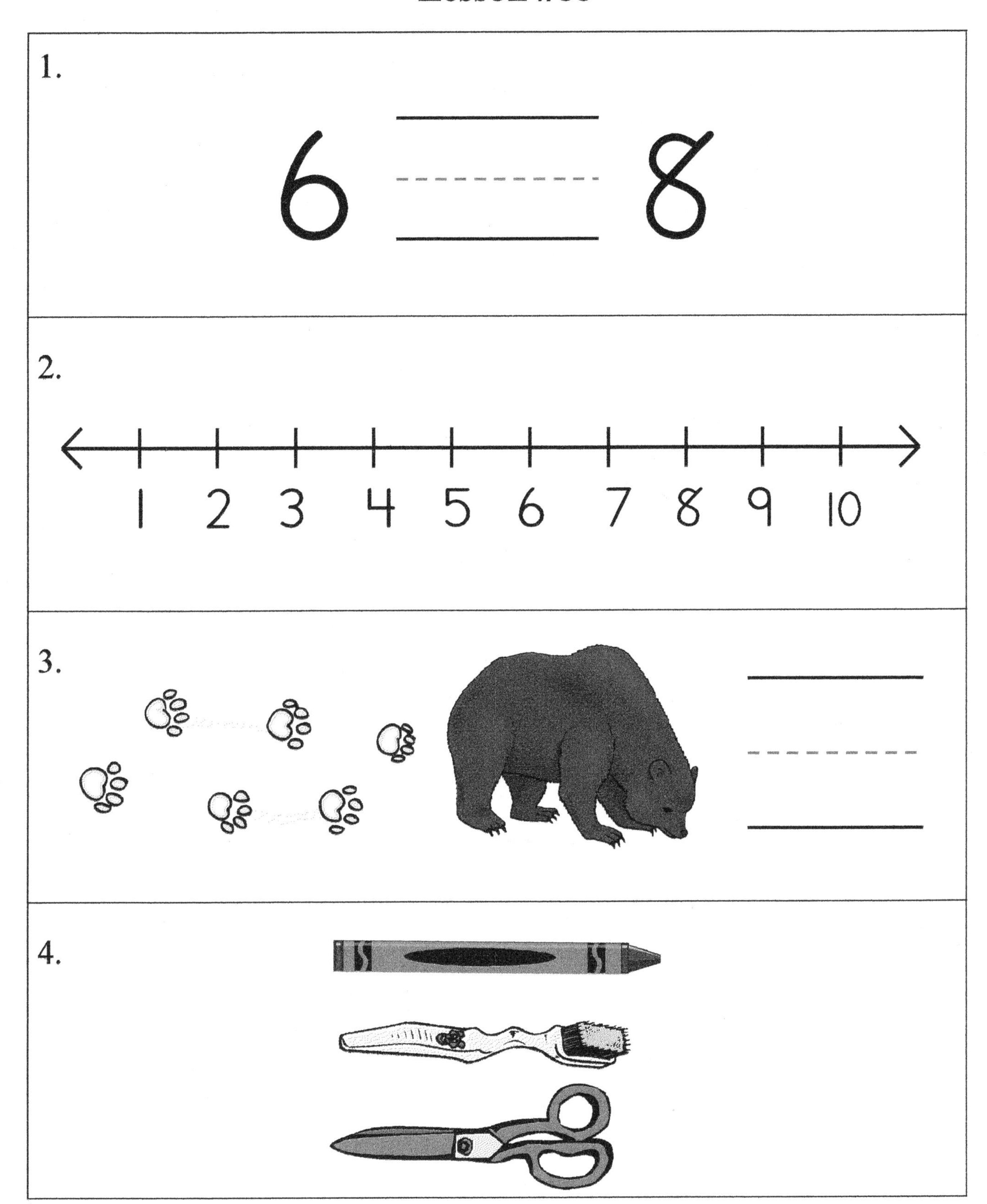

Directions:

1) Write the missing number.
2) Circle the number that comes before 8.
3) Count the animal tracks. Write the number.
4) Circle the object on the bottom.

Lesson #89

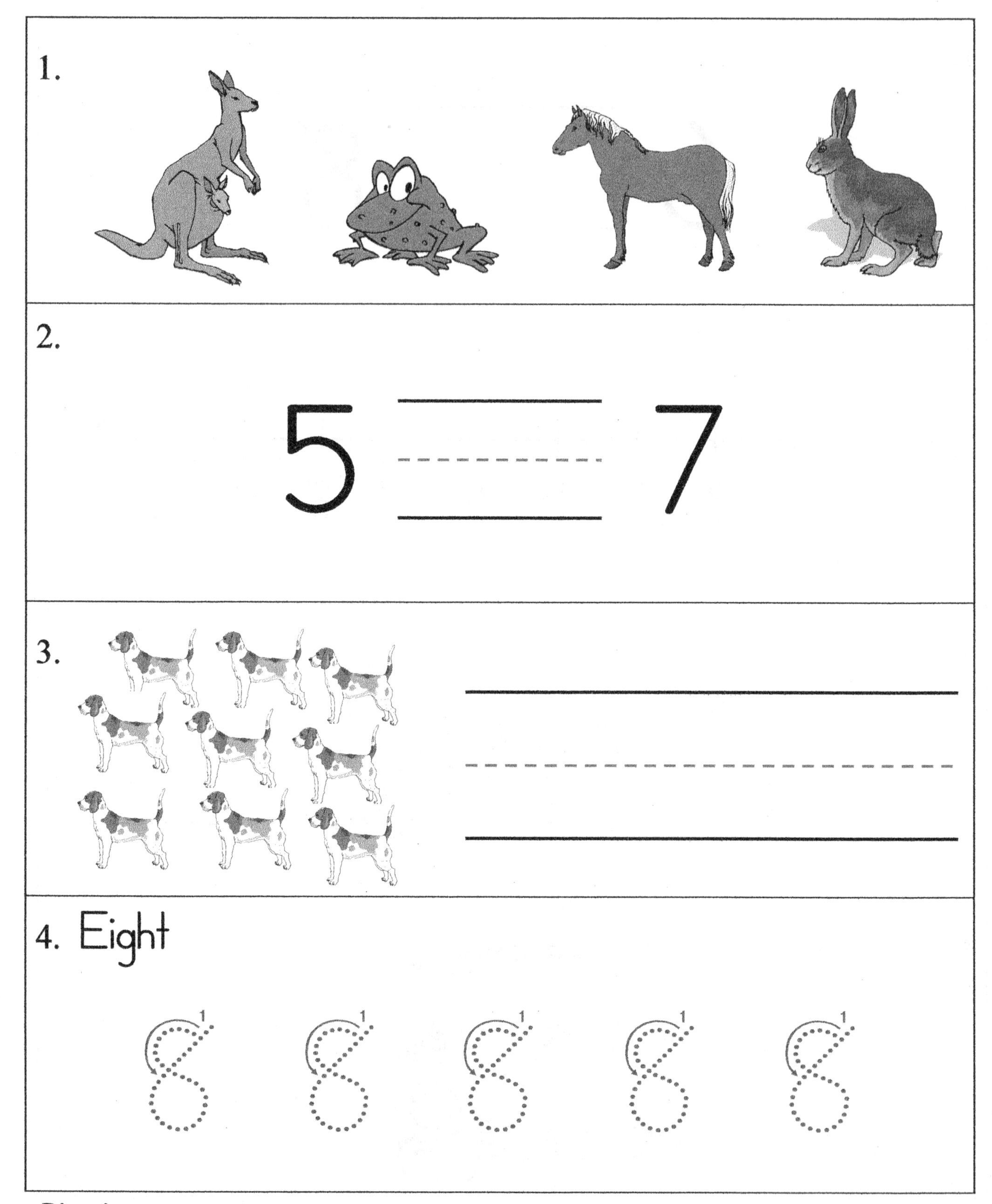

Directions:

1) Cross out the animal that doesn't belong.
2) Write the missing number.
3) Count the number of dogs. Write the number three times.
4) Trace the number.

Lesson #90

1.

2.

1 2 3 4 5 6 7 8 9 10

3.

4.

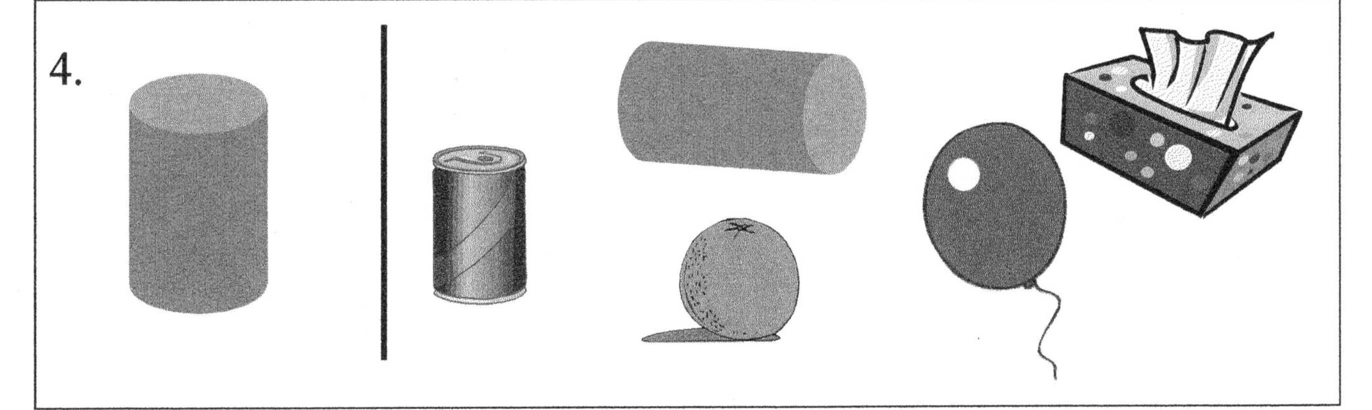

Directions:

1) Circle the animal that should come next.
2) Circle the number that comes before 7.
3) Write the missing number.
4) Circle the objects that are like the solid shape.

Lesson #91

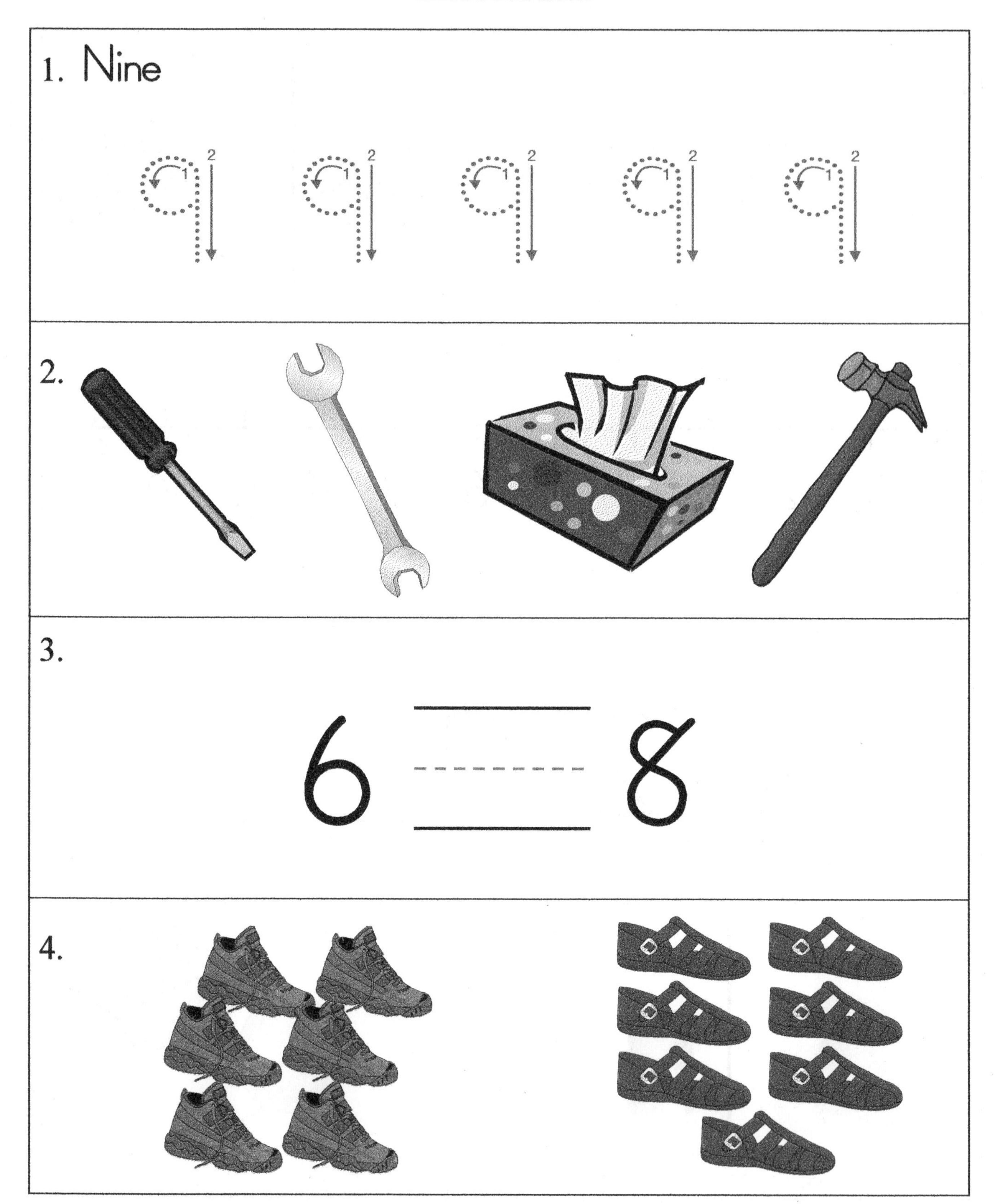

Directions:

1) Trace the number.
2) Cross out the picture that doesn't belong.
3) Write the missing number.
4) Count the number of shoes in each group. Circle the group with 7.

Lesson #92

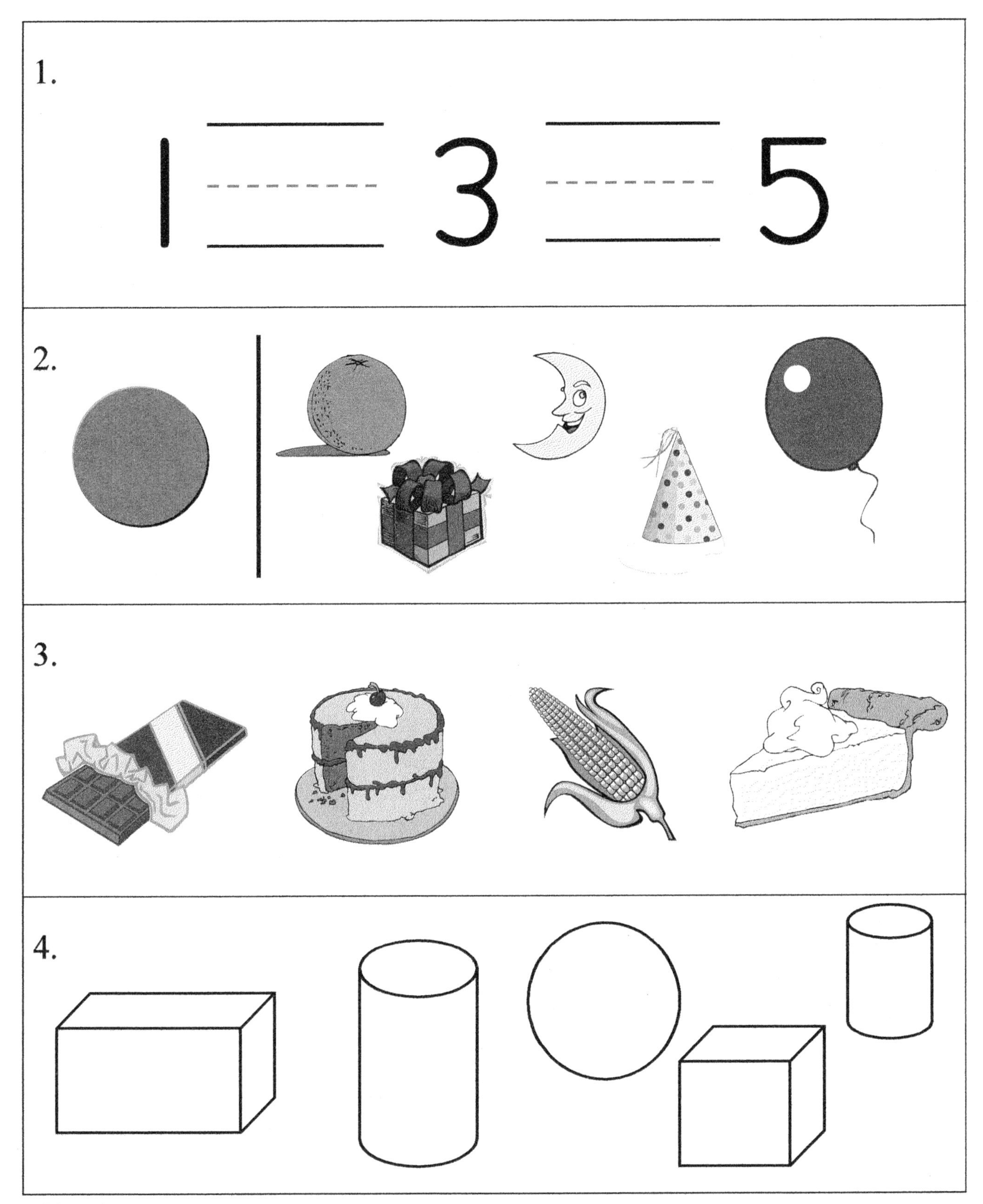

Directions:

1) Write the missing numbers.
2) Circle the objects that match the solid shape.
3) Cross out the picture that doesn't belong.
4) Color the shapes that can roll in green.

Lesson #93

1.

2.

3 ____ 5 ____ 7

3.

4. Ten

10 10 10 10 10

Directions:

1) Count the number of gifts. Write the number.
2) Write the missing numbers.
3) Circle the group where the shape belongs.
4) Trace the number.

Lesson #94

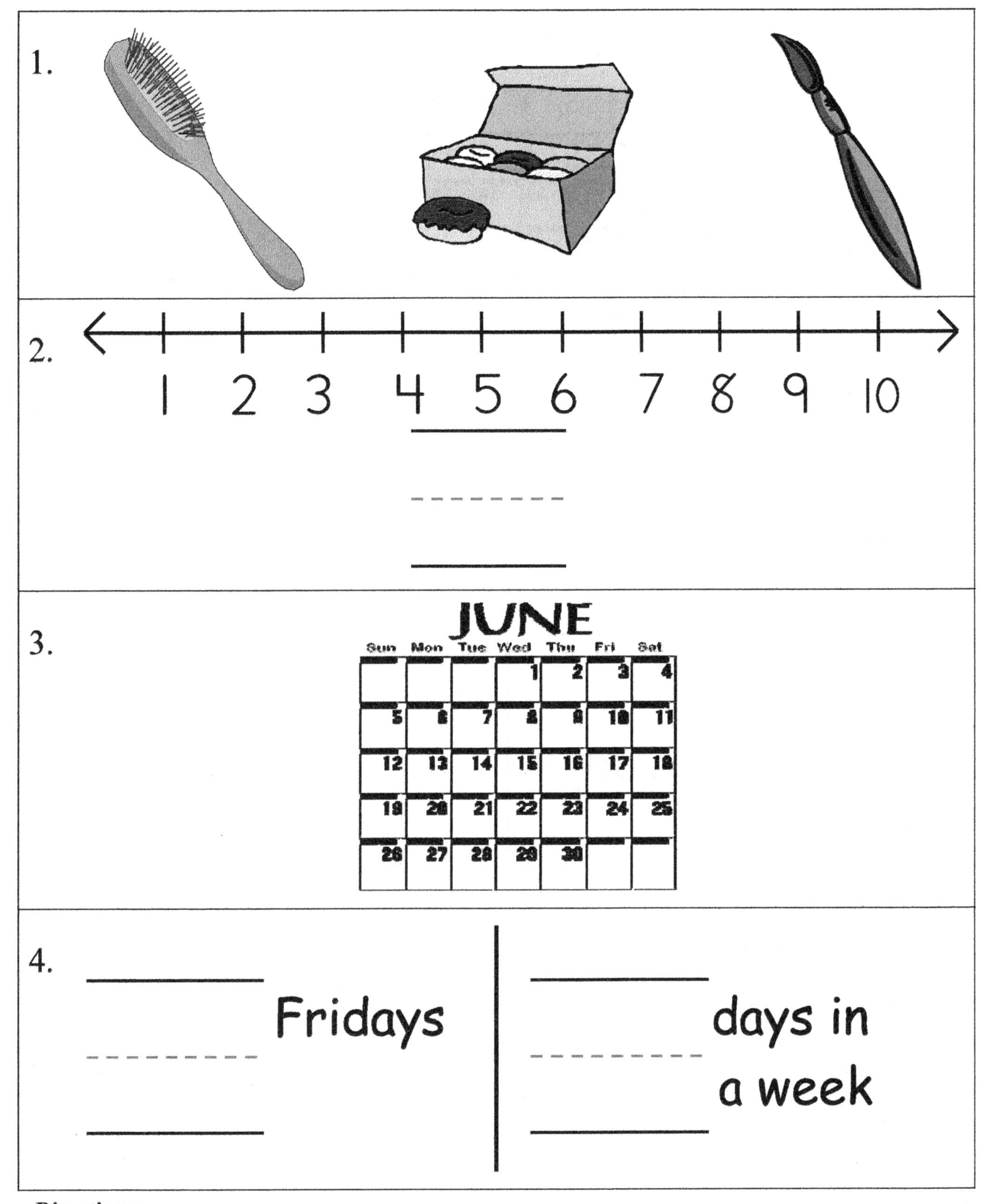

Directions:

1) Circle the object in the middle.
2) Write the number that comes after 5.
3) Circle every Friday in green.
4) Write the number of Fridays in the month and the number of days in one week.

Lesson #95

1.

JUNE

Sun	Mon	Tue	Wed	Thu	Fri	Sat
			1	2	3	4
5	6	7	8	9	10	11
12	13	14	15	16	17	18
19	20	21	22	23	24	25
26	27	28	29	30		

2. ______ Sundays | ______ days in a week

3. ______ Tuesdays | ______ Mondays

4. 5 ______ 7 ______ 9

Directions:

1) Circle every Sunday in red. Circle the first Tuesday in blue. Circle every Monday in green.
2) Write the number of Sundays. Count the number of days in one week.
3) Write the number of Tuesdays and Mondays.
4) Write the missing numbers.

Lesson #96

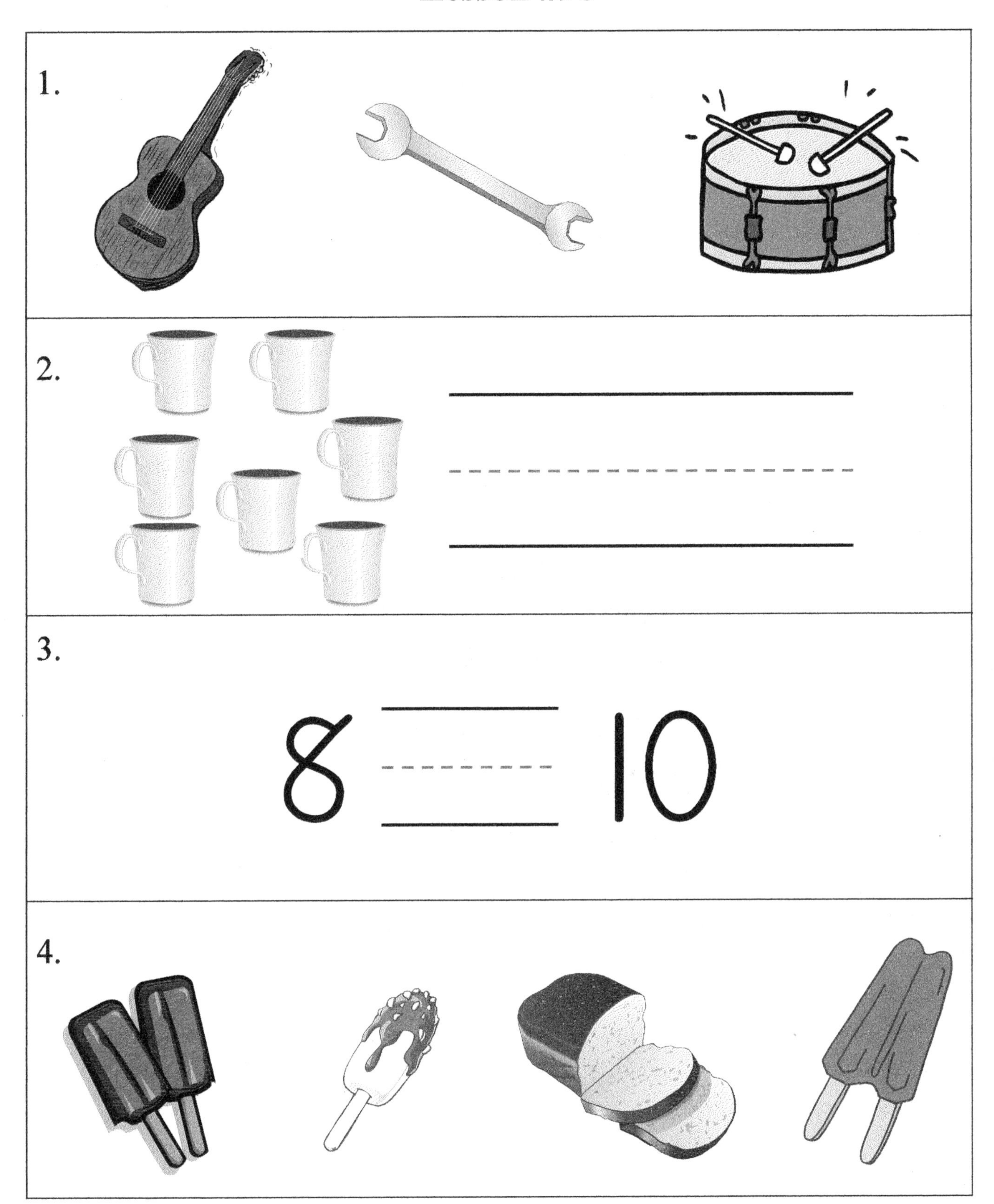

Directions:

1) Cross out the picture that doesn't belong.
2) Count the number of cups. Write the number.
3) Write the missing number.
4) Cross out the food that doesn't belong.

Lesson #97

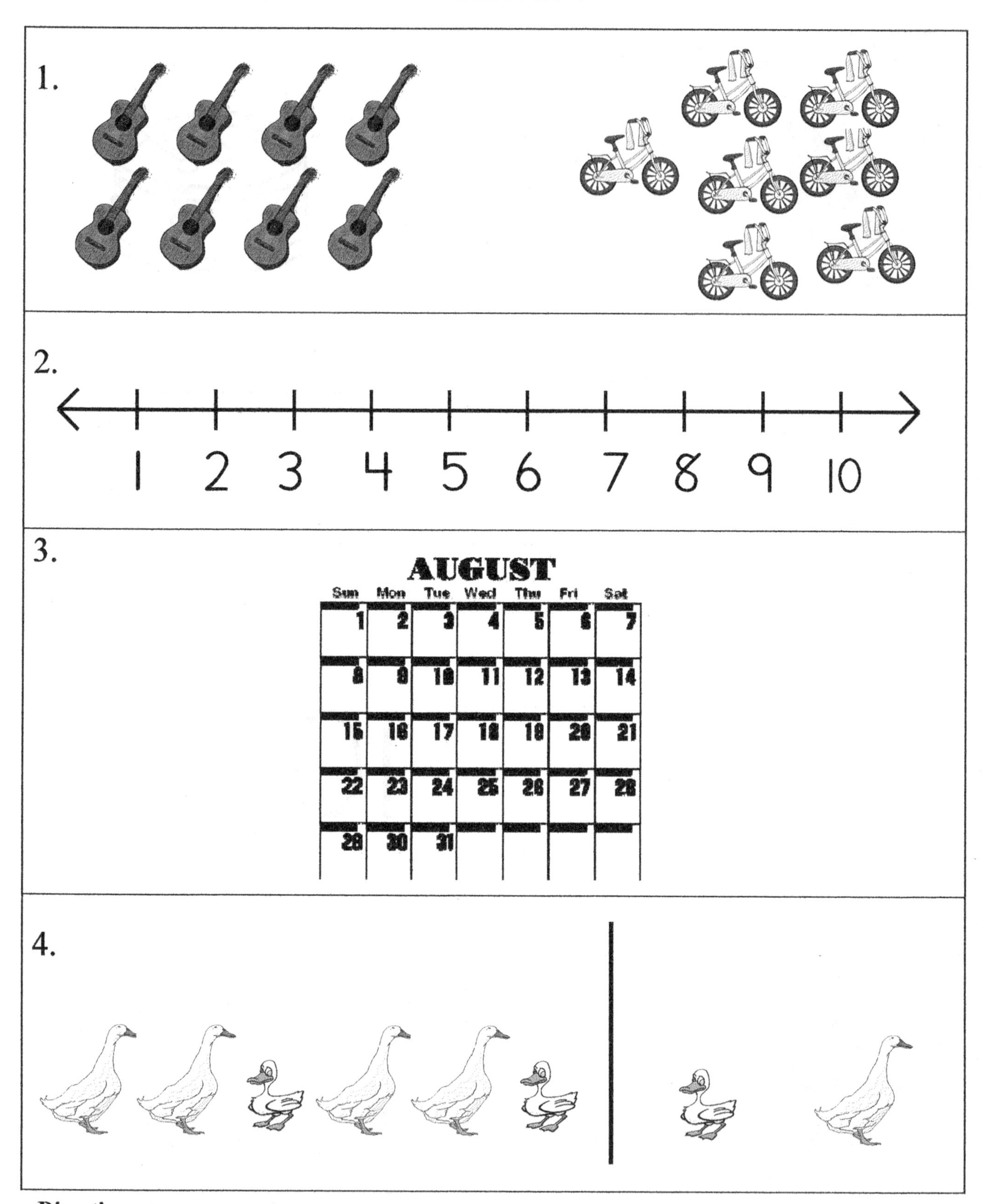

Directions:

1) Count the number in each group. Circle the group with 8.
2) Circle the number that comes before 6.
3) Circle every Sunday in orange.
4) Circle the one that should come next.

Lesson #98

Directions:

1) Count the number of chairs. Write the number.
2) Circle the picture that shows night.
3) Write the missing number.
4) Cross out the one that doesn't belong.

Lesson #99

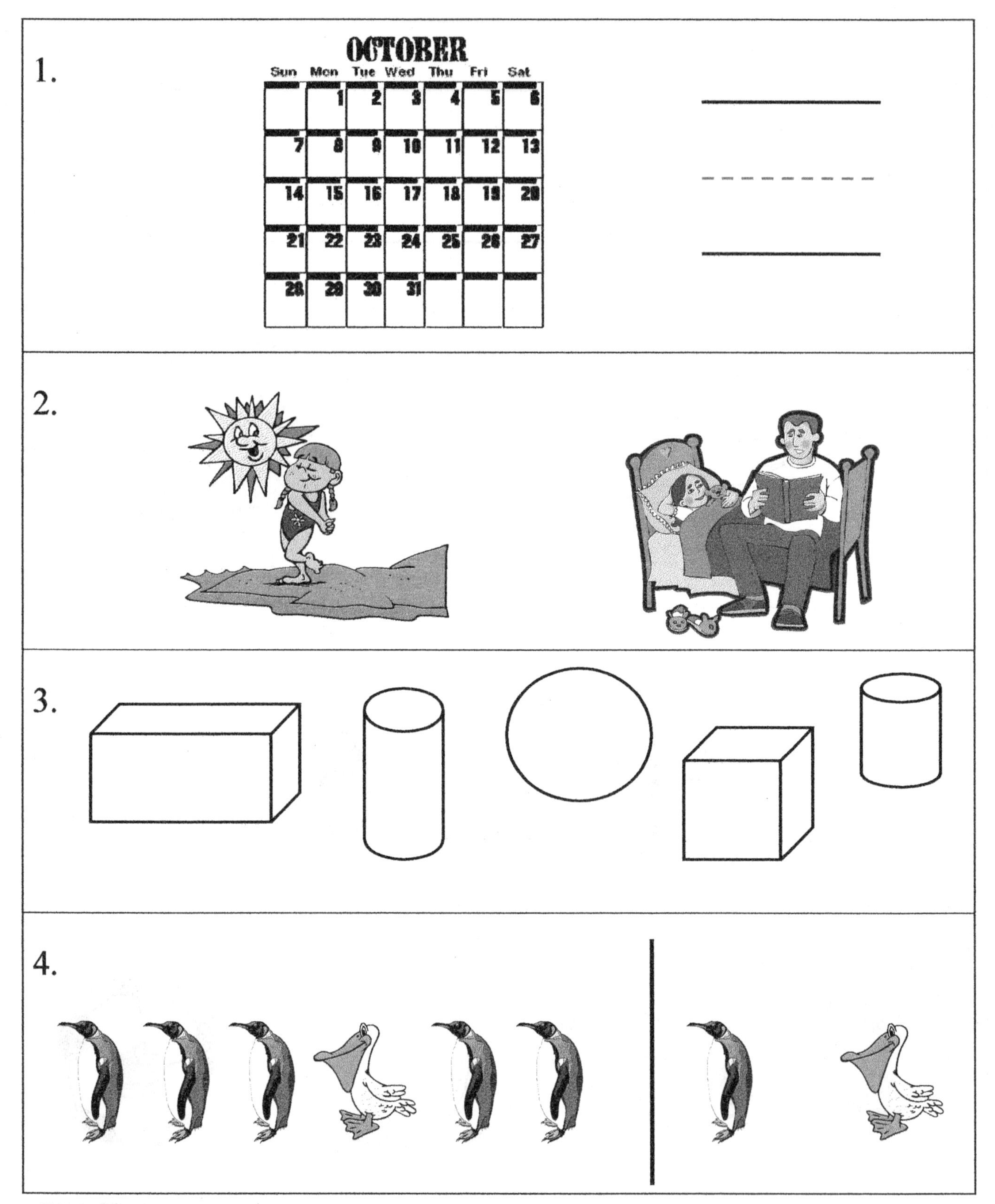

Directions:

1) Circle every Thursday in orange. Write the number of Thursdays.
2) Circle the picture that shows daytime.
3) Find the shapes that have corners. Color them red.
4) Circle the animal that should come next.

Lesson #100

1.

8 9 ____

2.

3.

4. Six

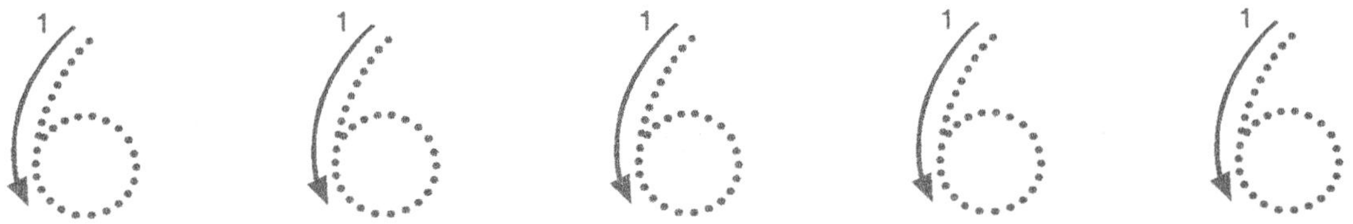

6 6 6 6 6

Directions:

1) Write the missing number.
2) Circle the picture that shows morning.
3) Count the number in each group. Circle the group with 9.
4) Trace the number.

Lesson #101

Directions:

1) Circle the picture that shows night.
2) Count the number of Saturdays in a month. Write the number.
3) Circle the activity that takes more time, watching a TV show or brushing your teeth.
4) Cross out the objects that are not musical instruments.

Lesson #102

1.

3 ____ 5 ____ 7

2.

3. Five

5 5 5 5 5

4.

Directions:

1) Write the missing numbers.
2) Count the number in each group. Circle the group with 10.
3) Trace the number.
4) Circle the shapes without corners.

Lesson #103

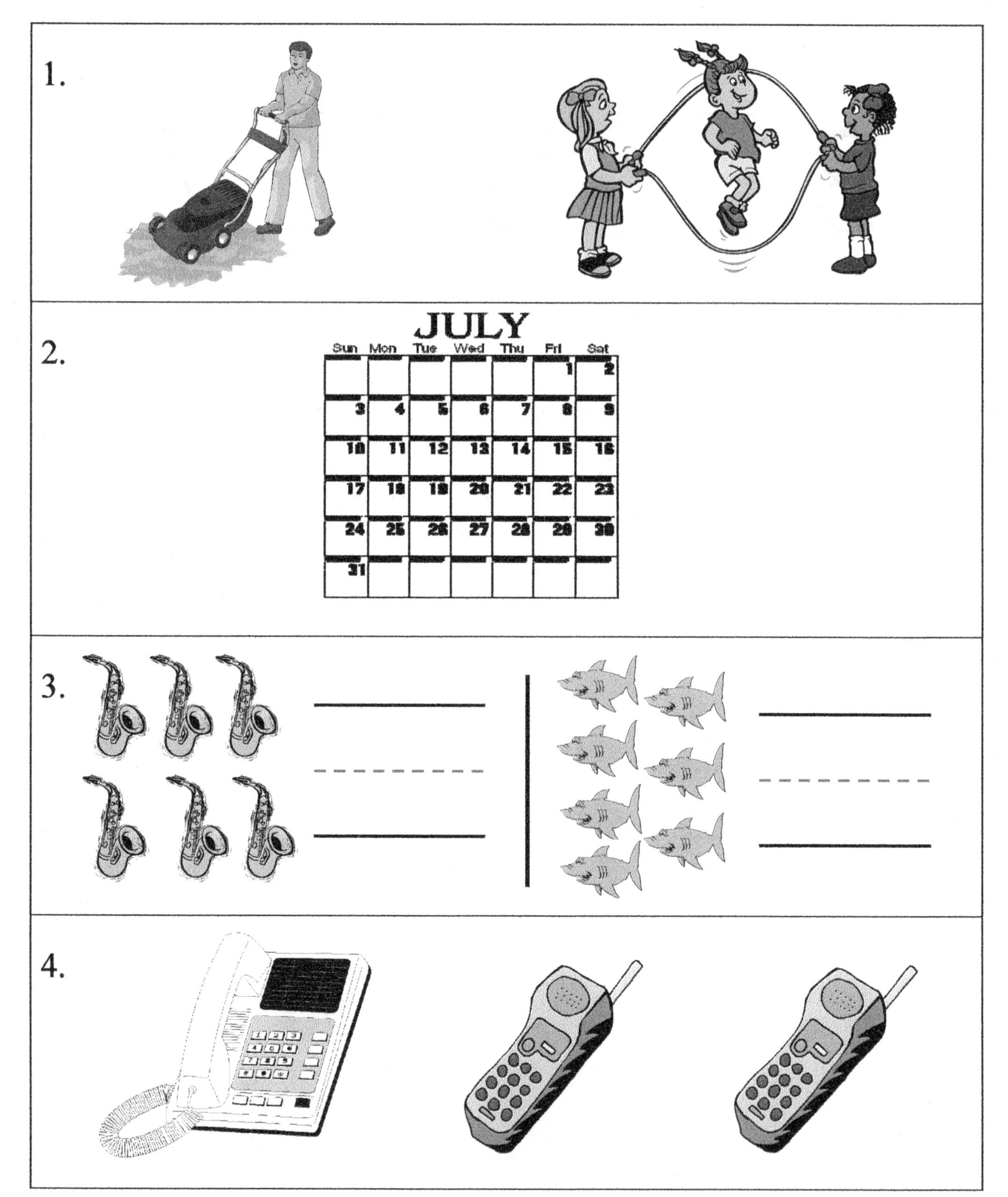

Directions:

1) Circle the activity that takes more time, cutting the grass or jumping rope.
2) Circle the first Friday in red. Circle every Wednesday in green.
3) Count the number in each group. Write the number.
4) Cross out the one that's different.

Lesson #104

1.

5 6 ___ 8 ___

2.

3.

4.

Directions:

1) Write the missing numbers.
2) Circle the picture that shows daytime.
3) Count the suns. Write the number.
4) Circle the one that should come next in the pattern.

Lesson #105

1.

____________ o'clock

2.

____________ o'clock

3.

4.

4 5 6 ____________

Directions:

1 – 2) Write the time shown on the clock.

3) Circle the activity that takes longer, riding the bus to school or kicking a ball.

4) Write the missing number.

Lesson #106

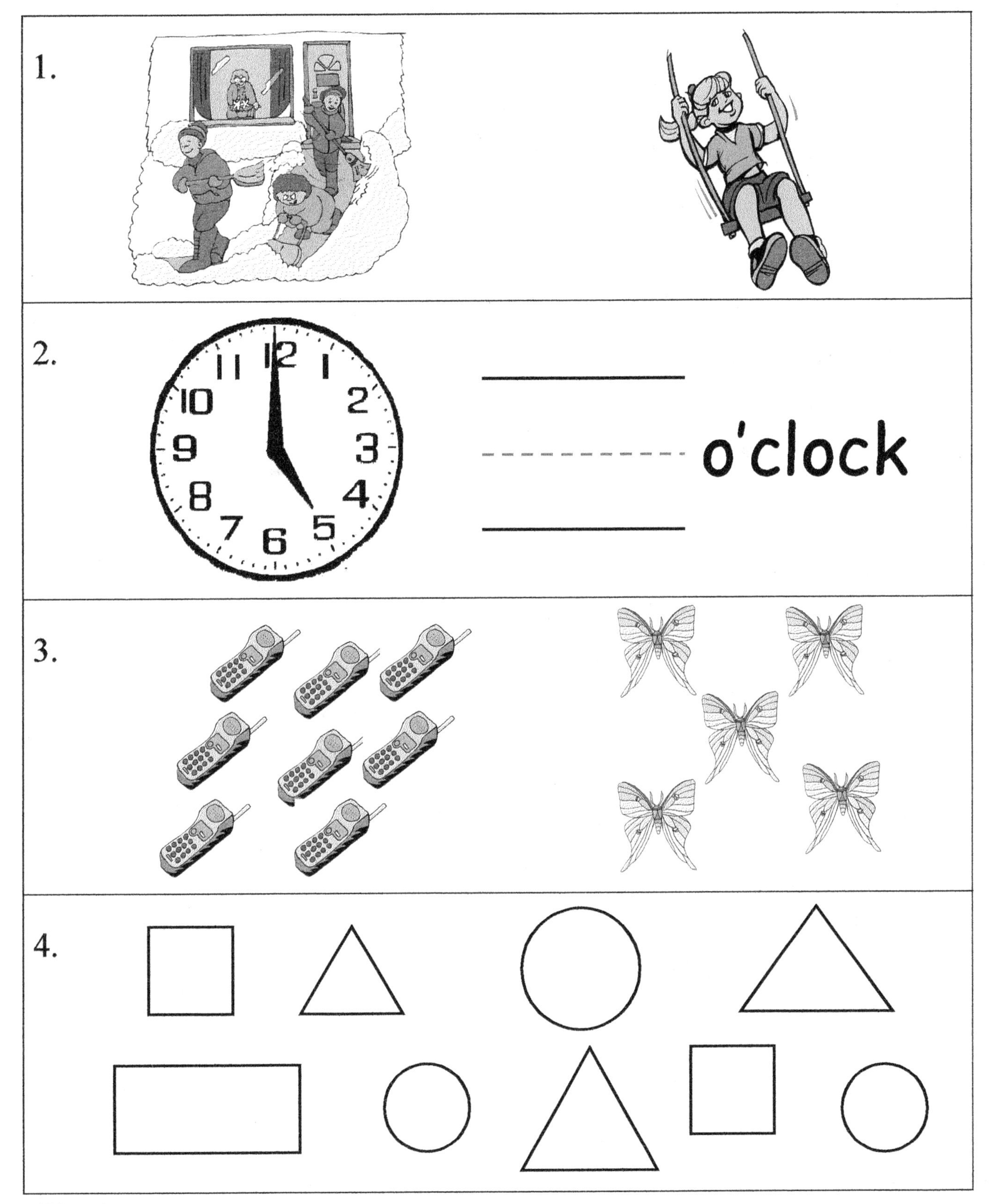

Directions:

1) Circle the activity that takes less time, shoveling snow or swinging.
2) Write the time shown on the clock.
3) Count the number in each group. Circle the group with 5.
4) Color the triangles in orange.

Lesson 107

1.

7 ____ 9

2.

____ o'clock

3.

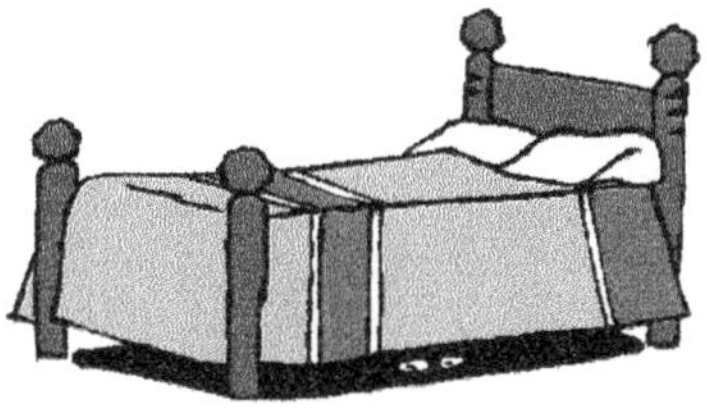

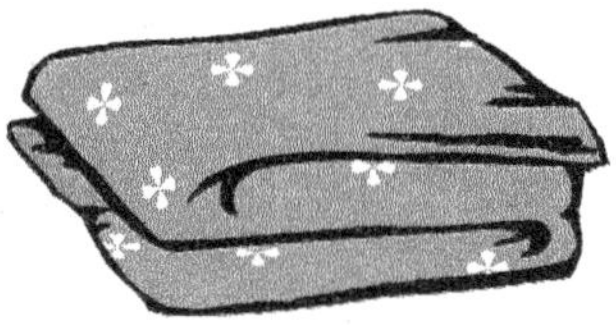

4.

Directions:

1) Write the missing number.
2) Write the time shown on the clock.
3) Cross out the one that doesn't belong.
4) Circle the activity that takes longer, brushing your teeth or practicing piano.

Lesson #108

1.

__________ o'clock

2.

AUGUST

Sun	Mon	Tue	Wed	Thu	Fri	Sat
1	2	3	4	5	6	7
8	9	10	11	12	13	14
15	16	17	18	19	20	21
22	23	24	25	26	27	28
29	30	31				

3. Four

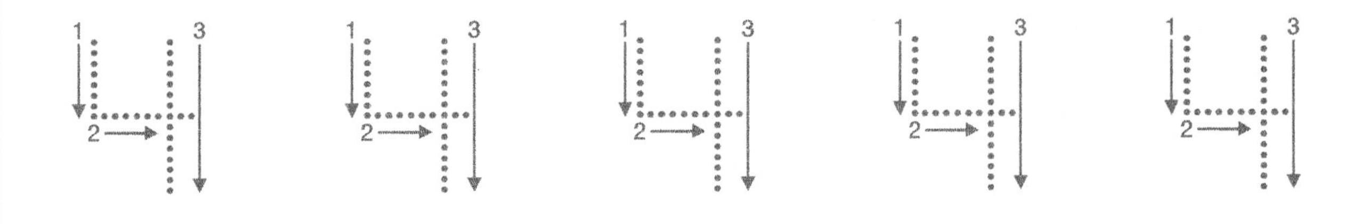

4.

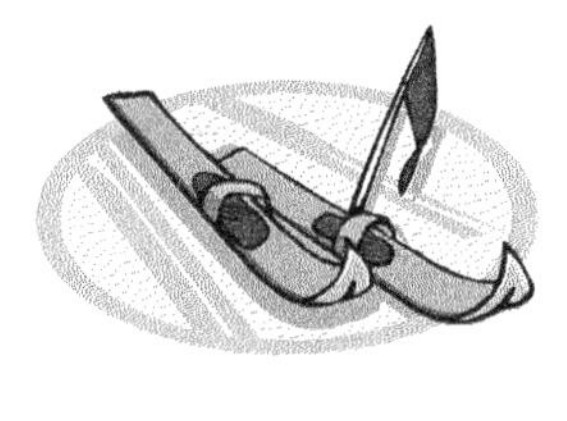

Directions:

1) Write the time shown on the clock.
2) Circle every Saturday in black. Circle the first Tuesday in blue.
3) Trace the number.
4) Cross out the picture that doesn't belong.

Lesson #109

1.

_______ o'clock

2.

2 3 ___ 5 6

3.

4.

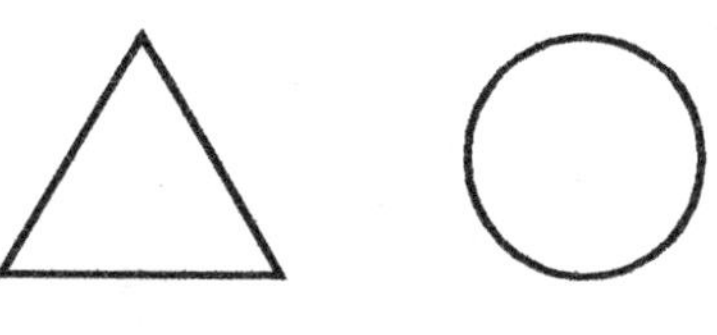 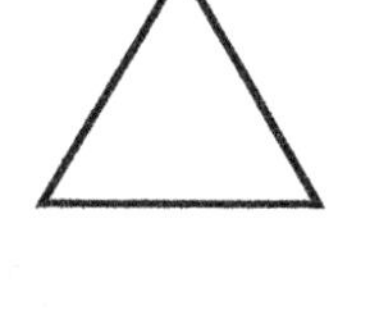  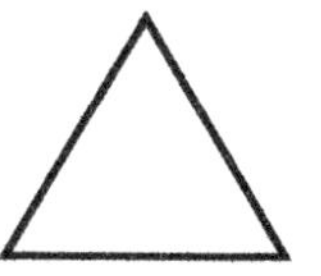

 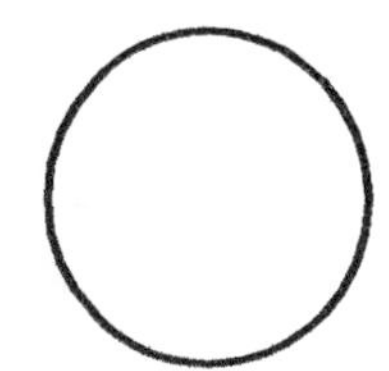  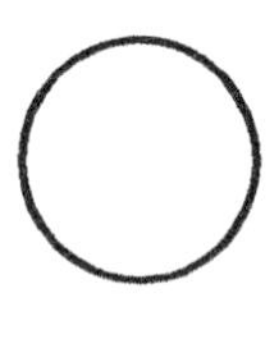

Directions:

1) Write the time shown on the clock.
2) Write the missing number.
3) Cross out the one that doesn't belong.
4) Color the rectangles green.

Lesson #110

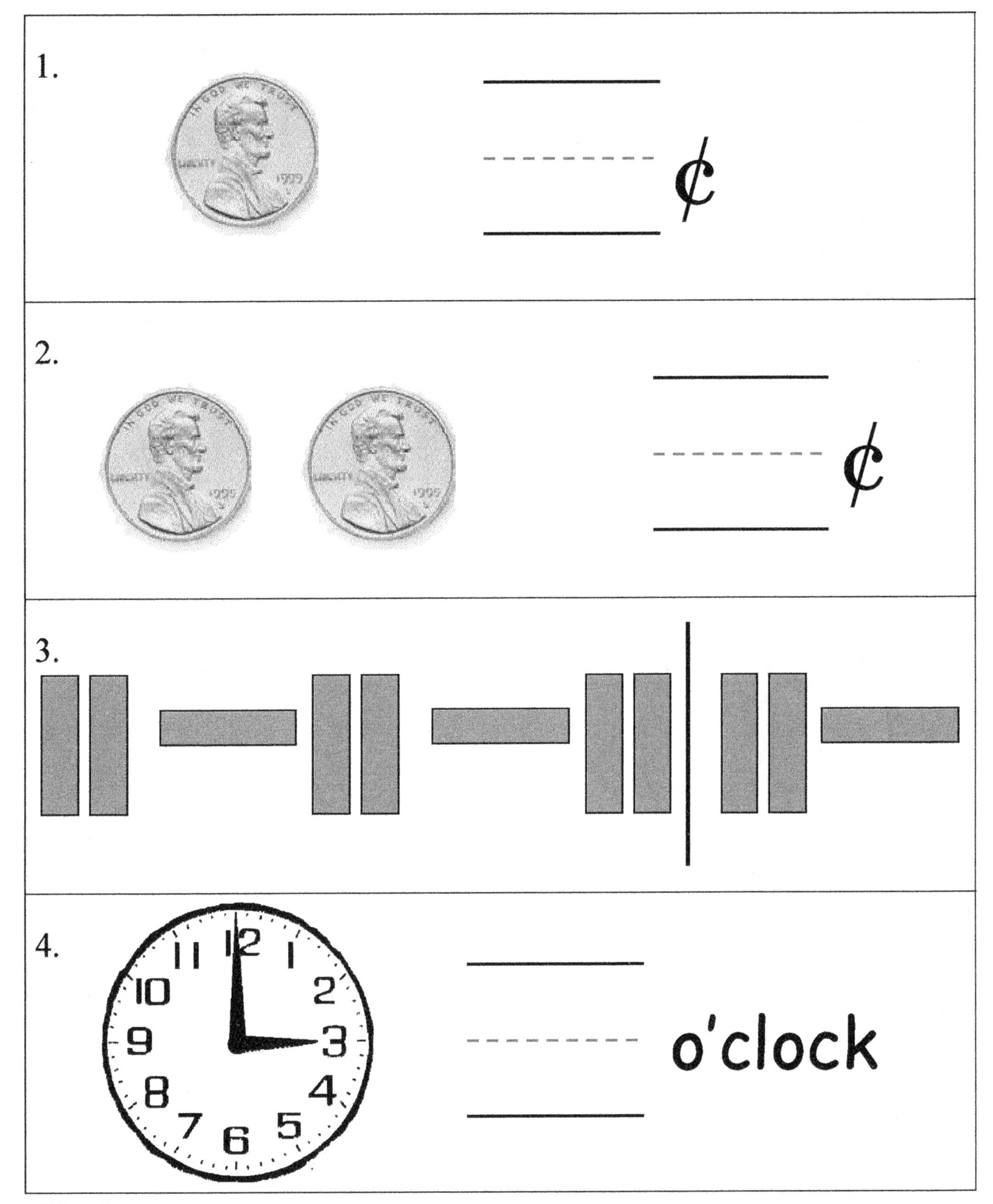

Directions:

1- 2) Write the number of cents.
3) Circle the shape that should come next in the pattern.
4) Write the time shown on the clock.

Lesson #111

1.

5 ______ 7 ______ 9

2.

3.

______ o'clock

4. Three

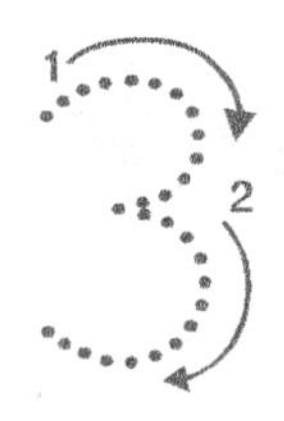
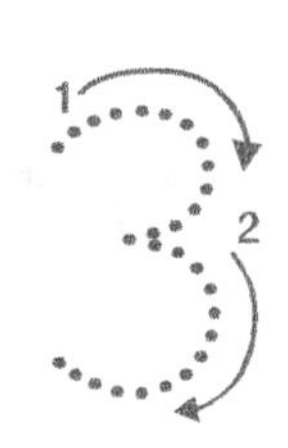
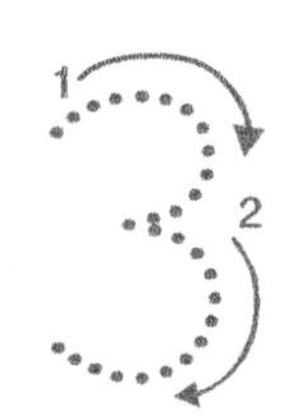
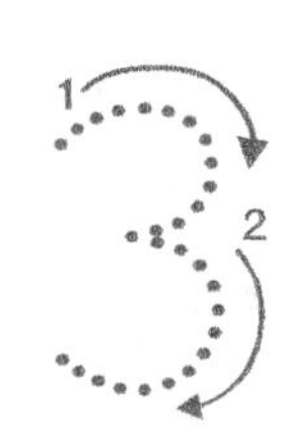
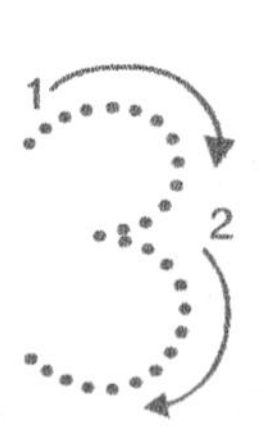

Directions:

1) Write the missing numbers.
2) Write the number of cents.
3) Write the time shown on the clock.
4) Trace the number.

Lesson #112

1.

______ ¢

2.

OCTOBER

Sun	Mon	Tue	Wed	Thu	Fri	Sat
	1	2	3	4	5	6
7	8	9	10	11	12	13
14	15	16	17	18	19	20
21	22	23	24	25	26	27
28	29	30	31			

3.

4.

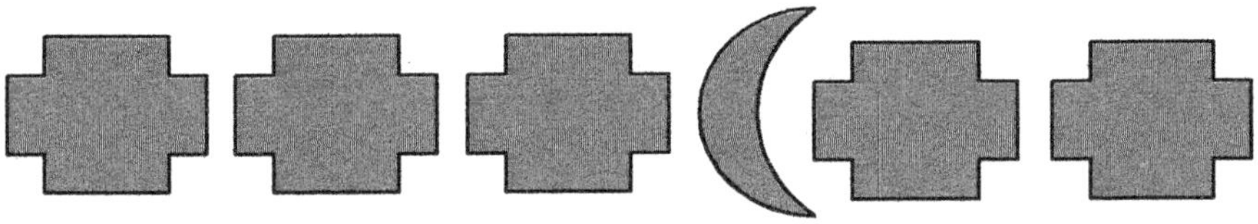

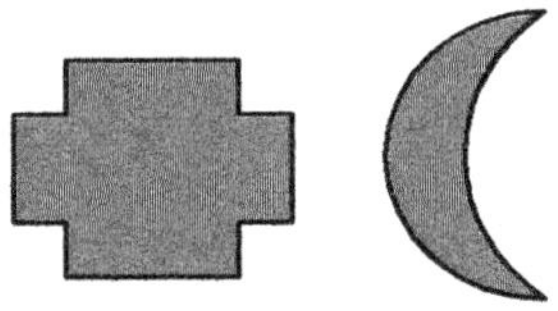

Directions:

1) Write the number of cents.
2) Circle every Friday in blue.
3) Circle the animal that doesn't belong.
4) Circle the shape that should come next in the pattern.

Lesson #113

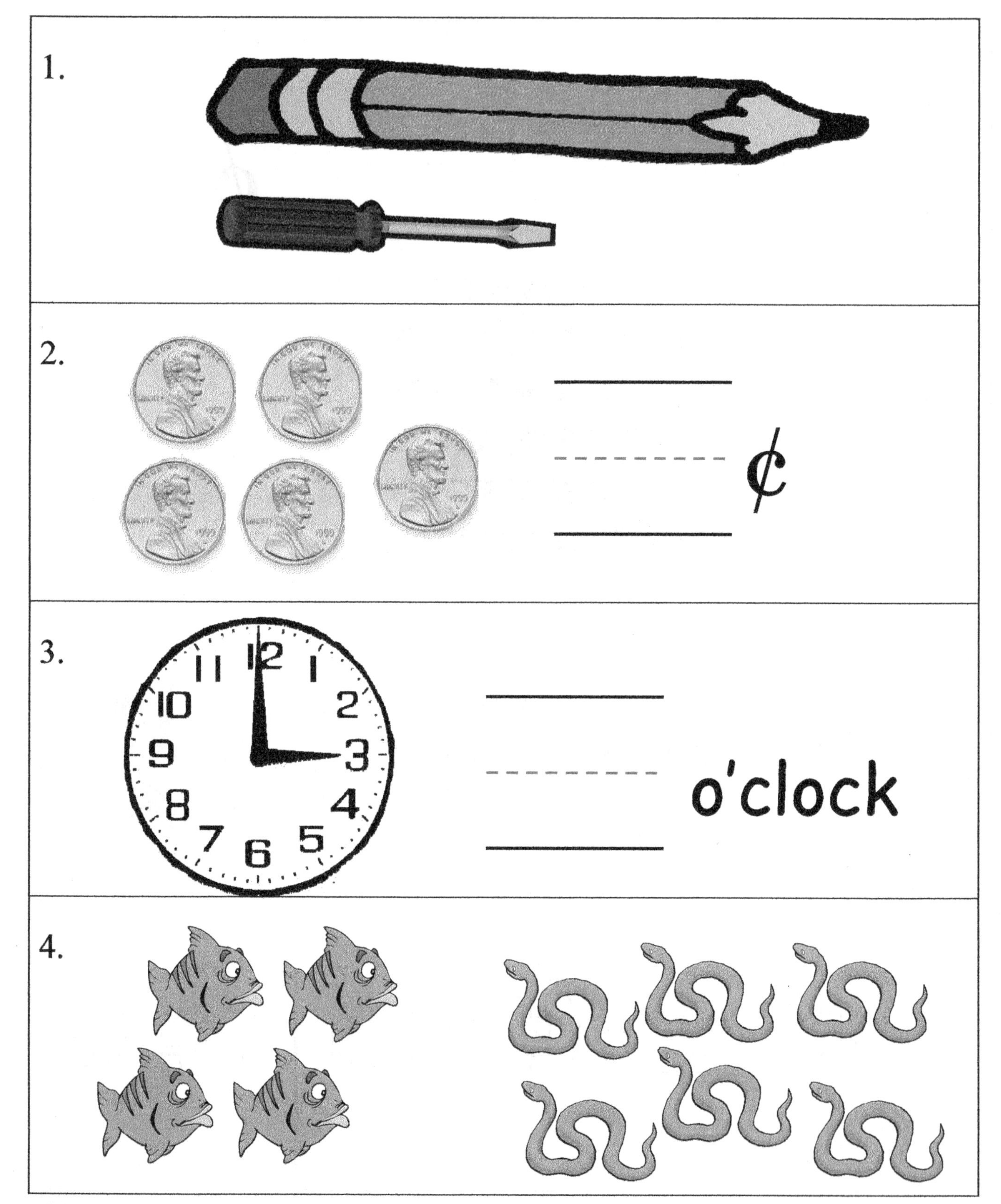

Directions:

1) Circle the object that is longer.
2) Write the number of cents.
3) Write the time shown on the clock.
4) Count the number in each group. Circle the group with fewer.

Lesson #114

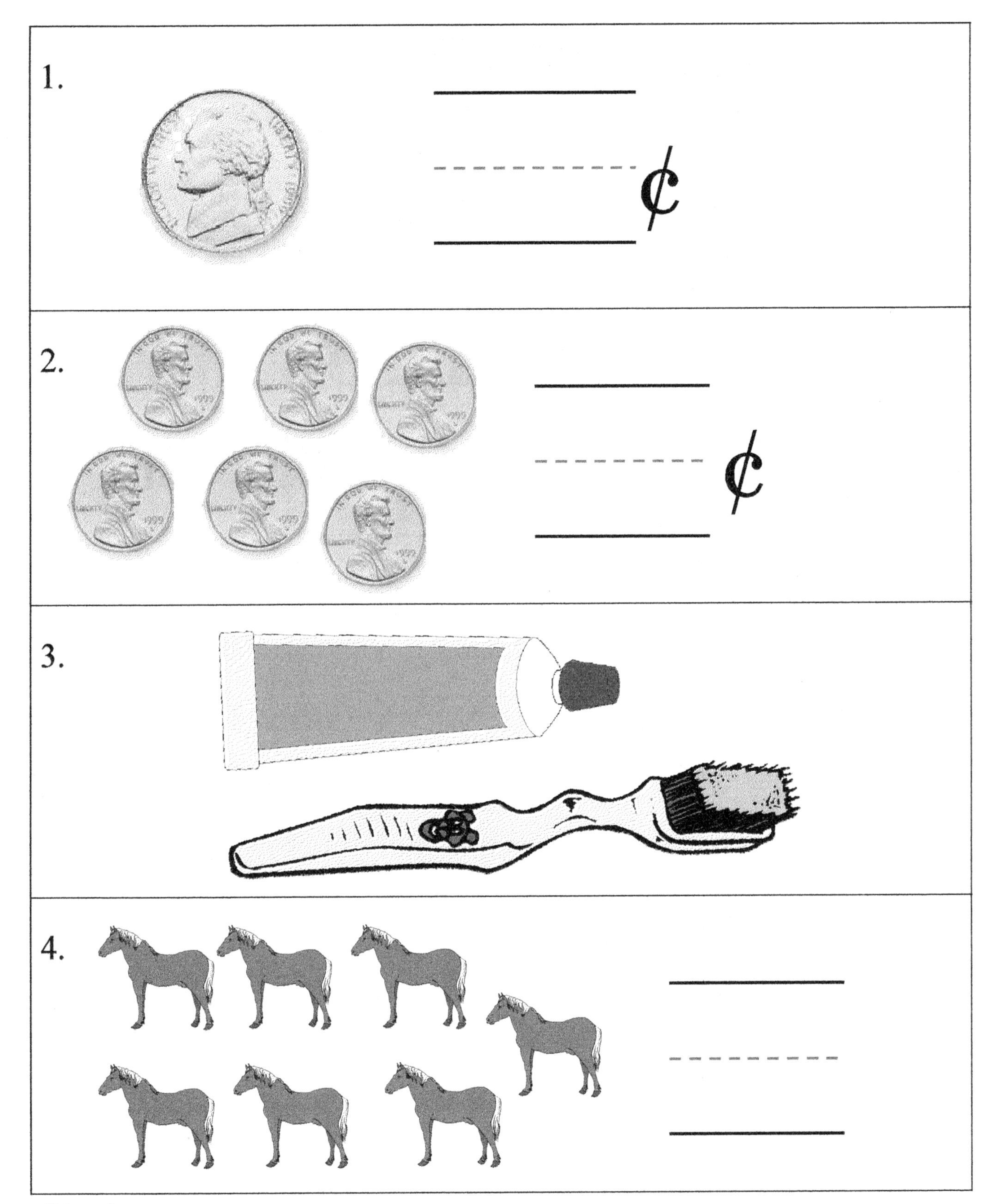

Directions:

1 – 2) Write the number of cents.
3) Circle the object that is shorter.
4) Count the number of horses. Write the number.

Lesson #115

Directions:

1) Underline the longer carrot.
2) Write the time shown on the clock.
3) Write the number of cents.
4) Circle the group where the shape belongs.

Lesson #116

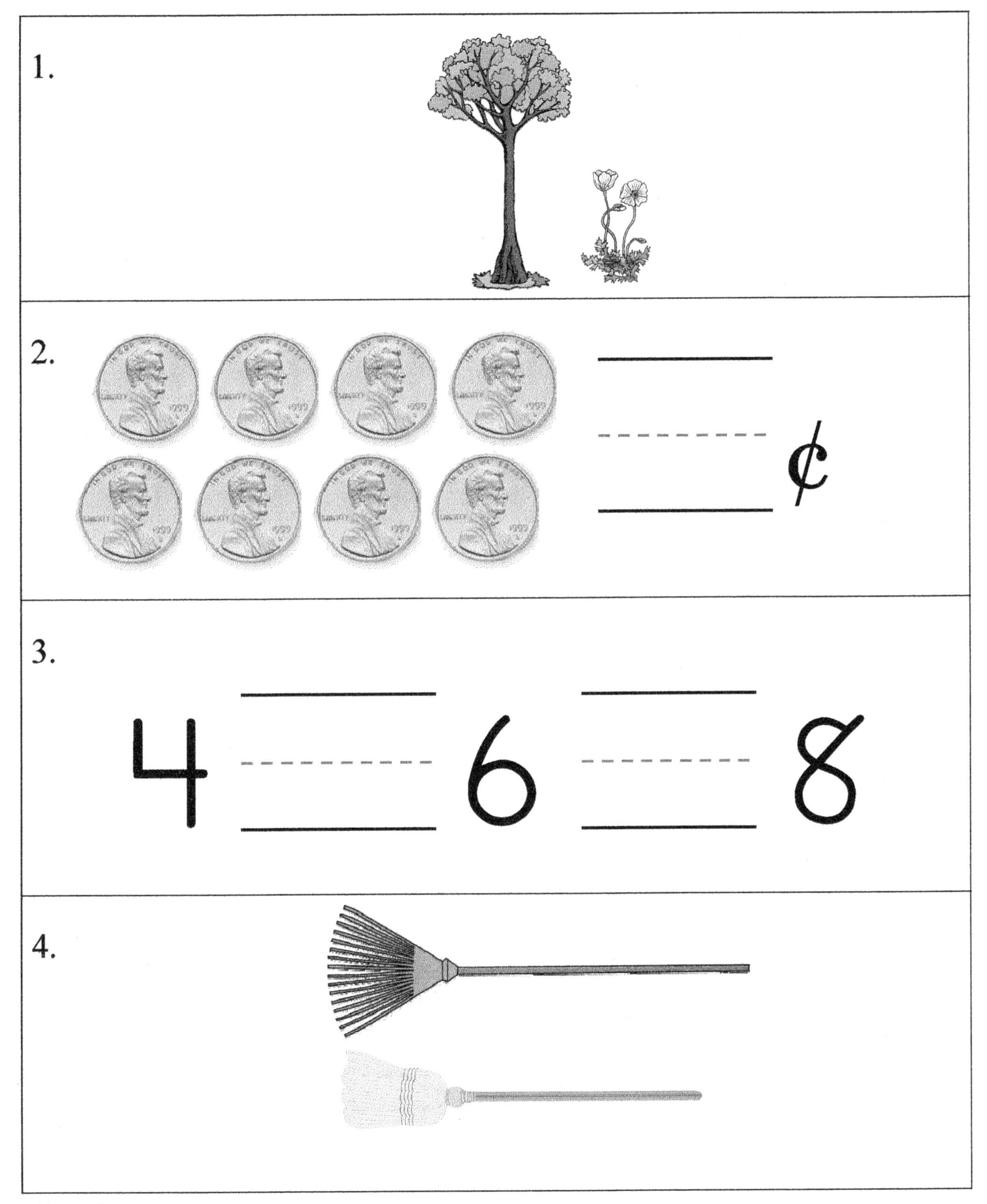

Directions:

1) Circle the object that is taller.
2) Write the number of cents.
3) Write the missing numbers.
4) Underline the shorter one.

Lesson #117

Directions:

1) Write the number of cents.
2) Circle the shorter one.
3) Write the time shown on the clock.
4) Color the squares orange.

Lesson #118

Directions:

1) Which one is heavier? Circle it.
2) Write the number of cents.
3) Circle the lighter one.
4) Write the missing number.

Lesson #119

1.

2.

3.

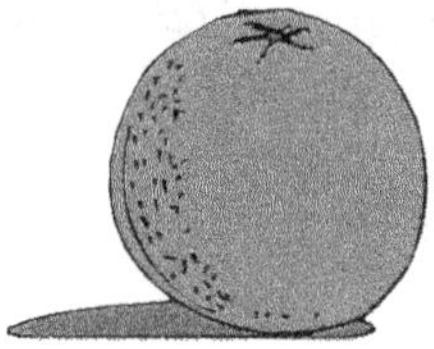

4.

______

- - - - - - - - - ¢

______

Directions:

1) Circle the longer one.
2) Count the number in each group. Circle the group with fewer.
3) Circle the heavier one.
4) Write the number of cents.

Lesson #120

1.

5 ____ 7 8 ____

2.

3.

____ o'clock

4.

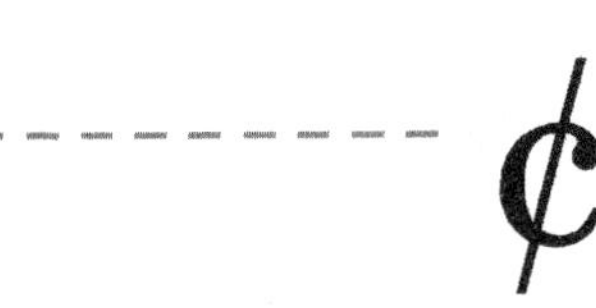

____ ¢

Directions:

1) Write the missing numbers.
2) Circle the lighter object.
3) Write the time shown on the clock.
4) Write the number of cents.

Lesson #121

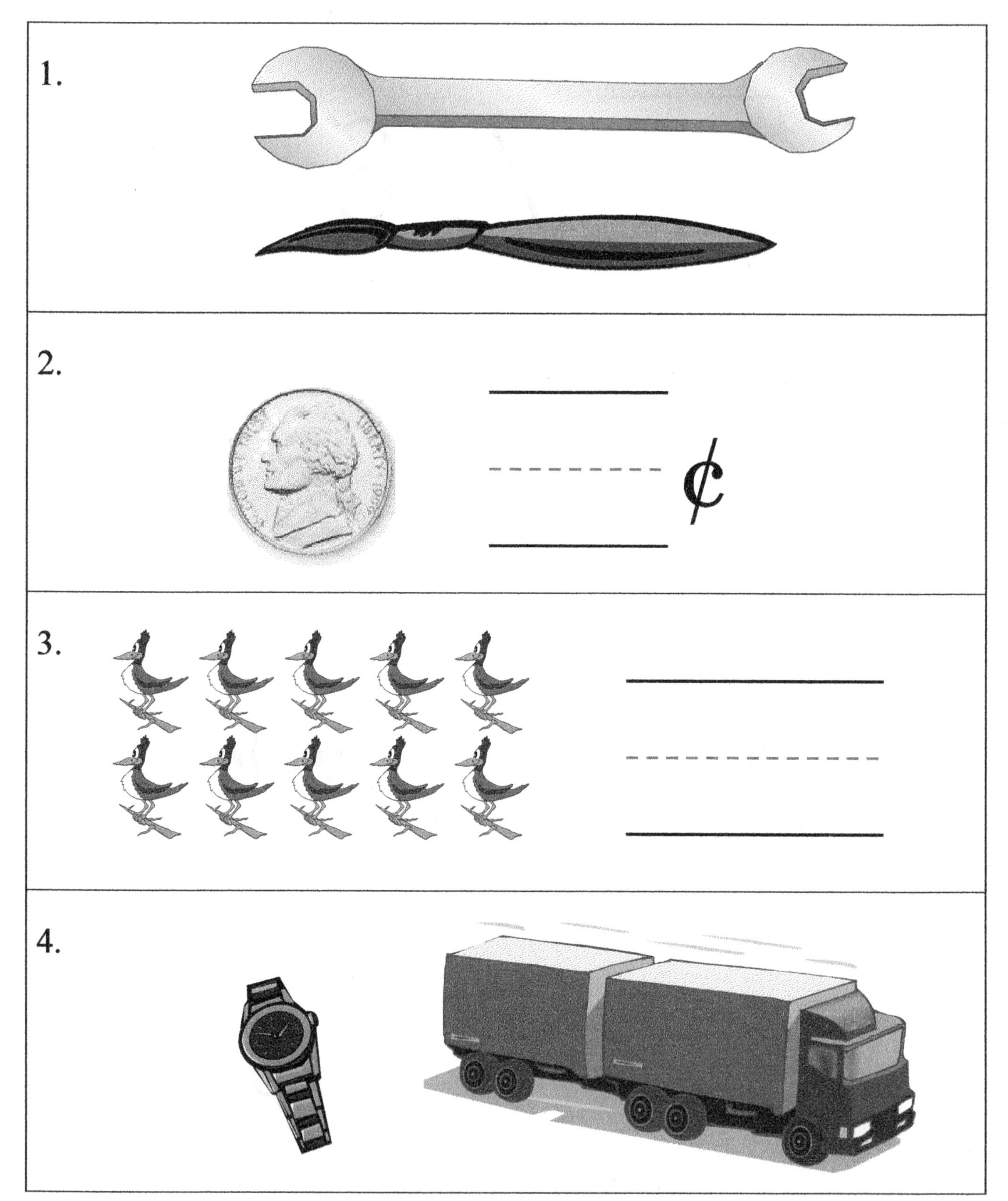

Directions:

1) Underline the shorter one.
2) Write the number of cents.
3) Count the number of birds. Write the number.
4) Circle the heavier one.

Lesson #122

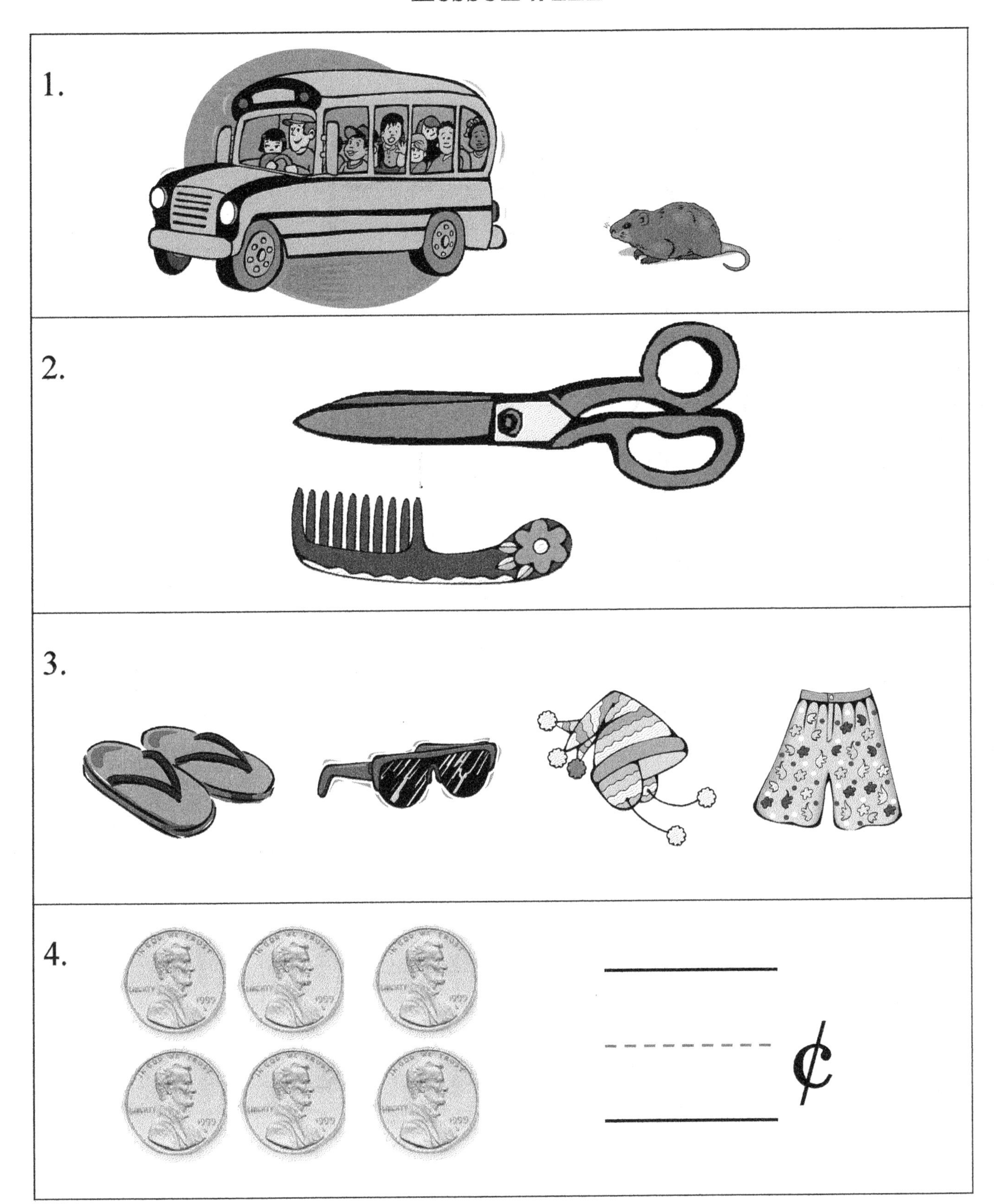

Directions:

1) Circle the one that is taller.
2) Circle the longer one.
3) Cross out the one that doesn't belong.
4) Write the number of cents.

Lesson #123

1.

7 8 9 ______

2.

______ o'clock

3.

4.

______ ¢

Directions:

1) Write the missing number.
2) Write the time shown on the clock.
3) Color the triangles blue.
4) Write the number of cents.

Lesson #124

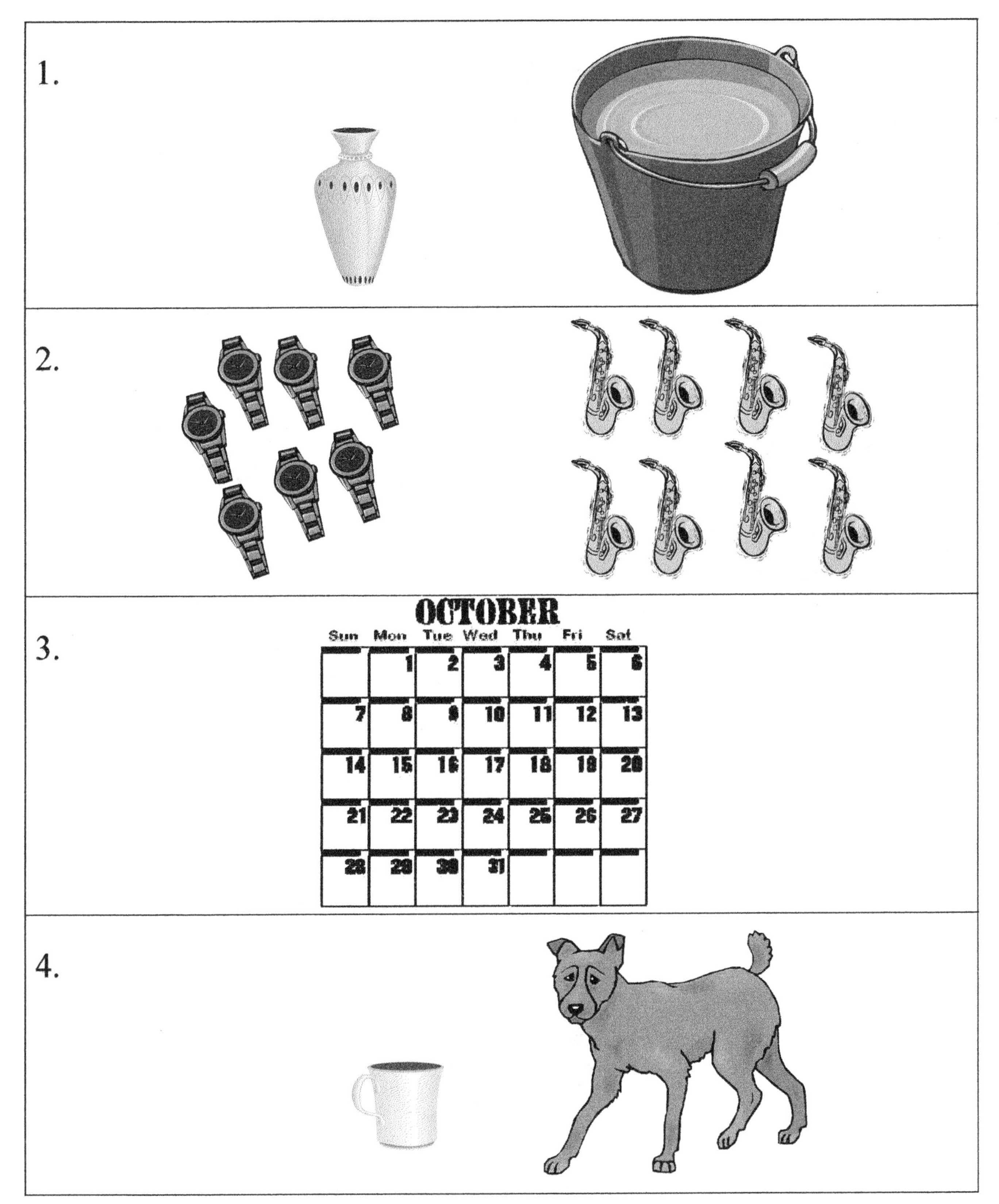

Directions:

1) Circle the one that holds more.
2) Circle the group with 8 in it.
3) Circle every Monday in red.
4) Cross out the lighter one.

Lesson #125

1.

2.

11 12 1
10 2
9 3
8 4
7 6 5

o'clock

3.

4.

4 5 ____

Directions:

1) Circle the nickel.
2) Write the time shown on the clock.
3) Circle the one that holds more water.
4) Write the missing number.

Lesson #126

1.

2.

3.

4.

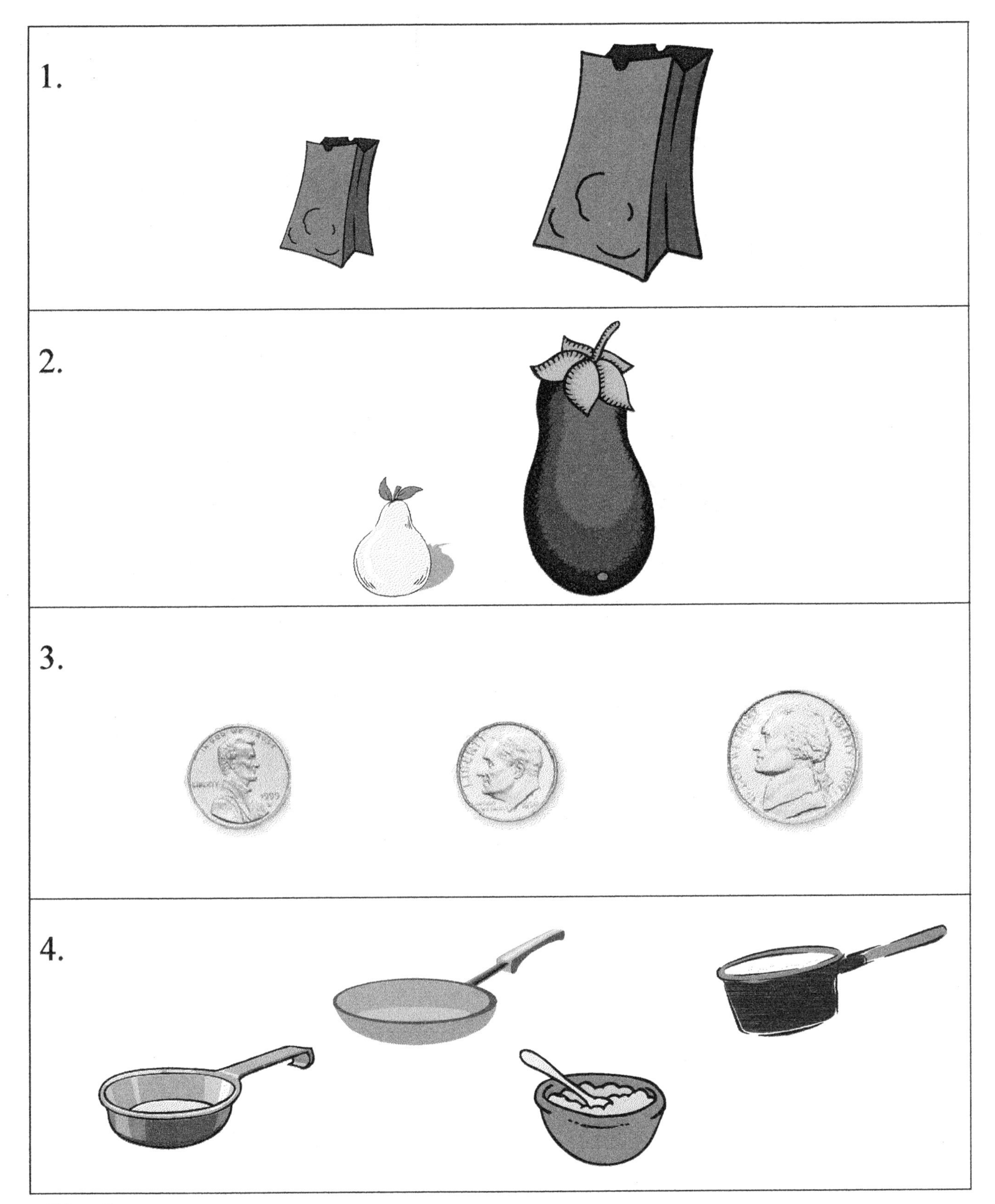

Directions:

1) Circle the bag that holds more.
2) Cross out the one that is shorter.
3) Circle the dime.
4) Cross out the one that doesn't belong.

Lesson #127

1. ______ o'clock

2.

3. ______ ¢

4.

Directions:

1) Write the time shown on the clock.
2) Circle the one that holds less.
3) Write the number of cents.
4) Circle the one that is heavier.

Lesson #128

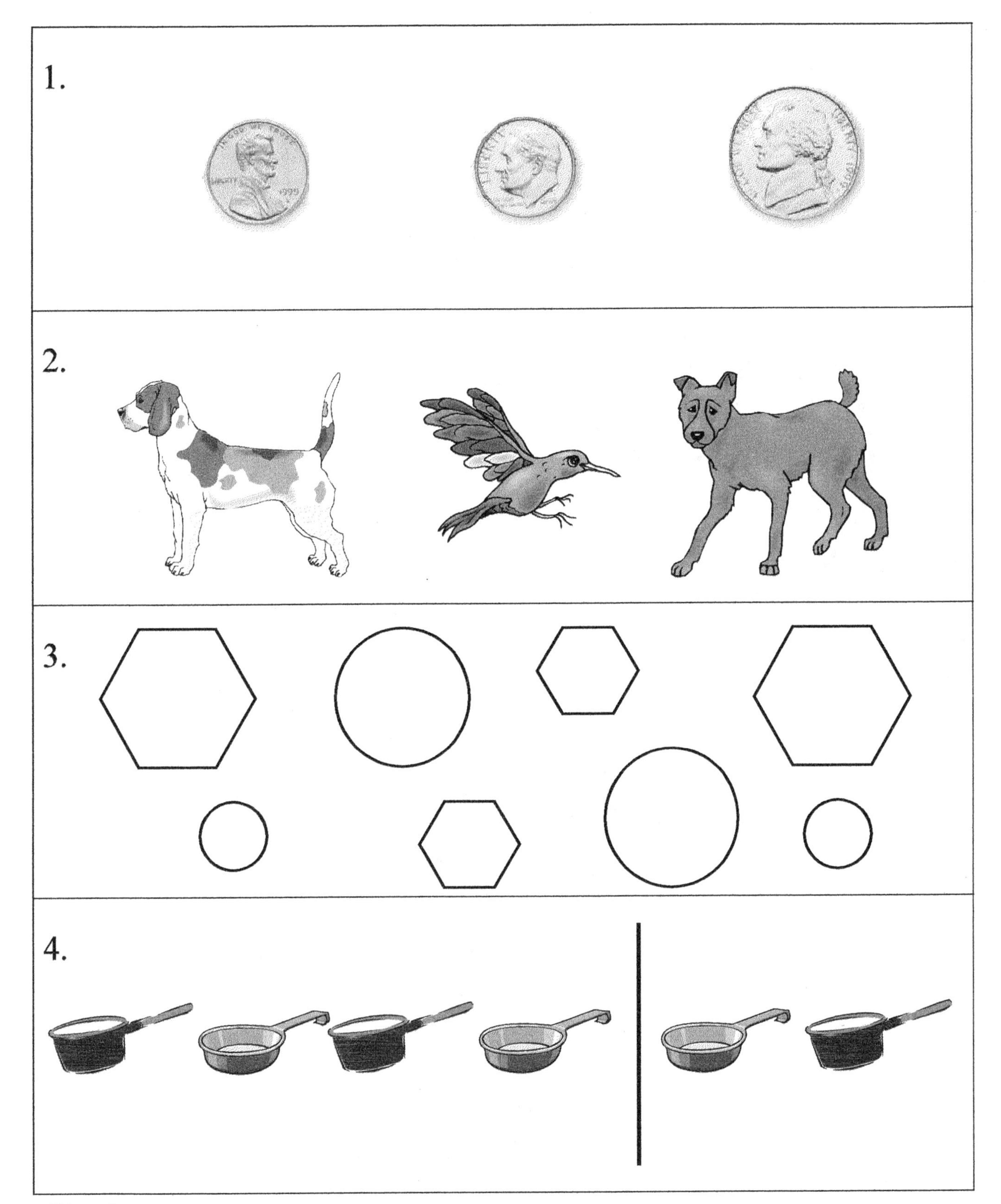

Directions:

1) Circle the penny.
2) Circle the two that are about the same weight.
3) Color the circles orange.
4) Circle the object that should come next.

Lesson #129

Directions:

1) Circle the shorter one.
2) Write the time shown on the clock.
3) Circle the pot that holds less.
4) Cross out the two that weigh about the same.

Lesson #130

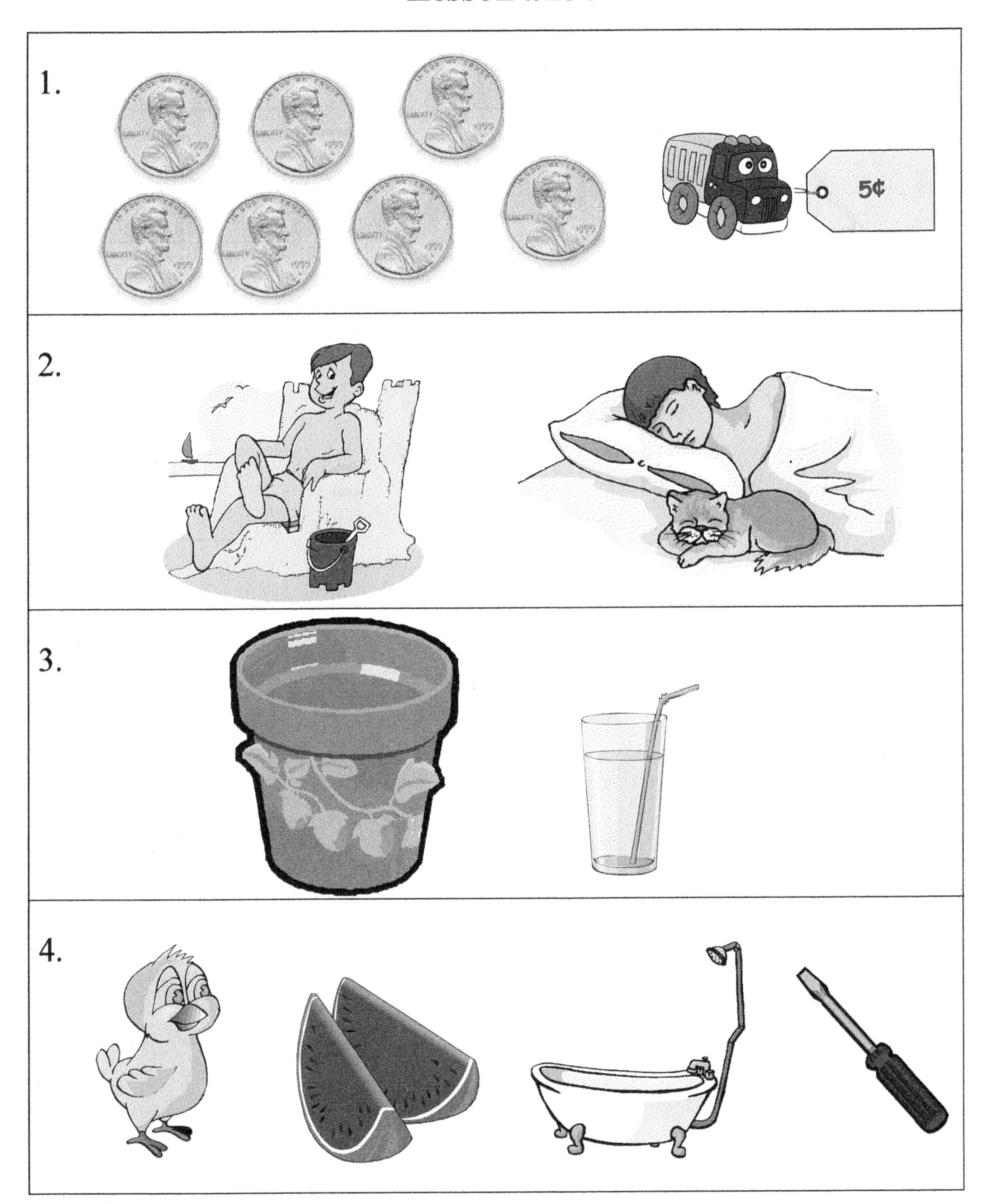

Directions:

1) Circle the pennies used to buy the toy.
2) Circle the picture that shows daytime.
3) Cross out the one that holds more.
4) Cross out the picture after the watermelon.

Lesson #131

1.

2 3 ___ 5 ___

2.

1 + 2 = ___

3.

7¢

4.

Directions:

1) Write the missing numbers.
2) Write how many ducks in all.
3) Circle the pennies used to buy the toy.
4) Cross out the one that holds less.

Lesson #132

Directions:

1) Write how many ice cream cones in all.
2) Write the time shown on the clock.
3) Circle the activity that takes less time, blowing a bubble or playing a basketball game.
4) Circle the two that weigh about the same.

Lesson #133

1.

6 ____ 8

2.

1 + 1 = ____

3.

4.

____ ¢

Directions:

1) Write the missing number.
2) How many hamburgers in all? Write the number.
3) Cross out the picture before the snake.
4) Write how many cents.

Lesson #134

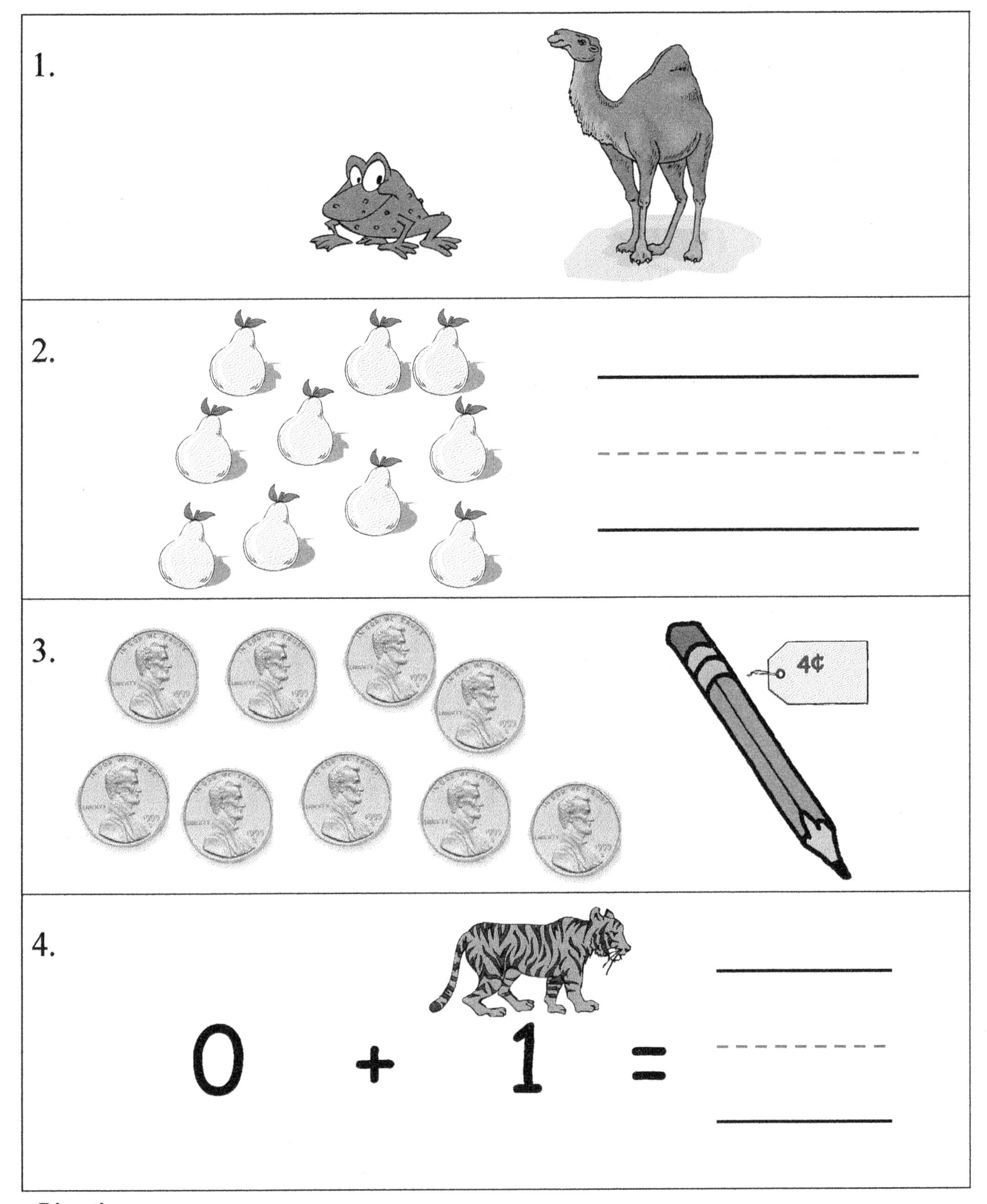

Directions:

1) Underline the taller one.
2) Count the number of pears. Write the number.
3) Circle the number of pennies used to buy the pencil.
4) Write how many tigers in all.

Lesson #135

Directions:

1) Write the time shown on the clock.
2) Write how many fish in all.
3) Circle the one that holds more.
4) Circle the shape that should come next.

Lesson #136

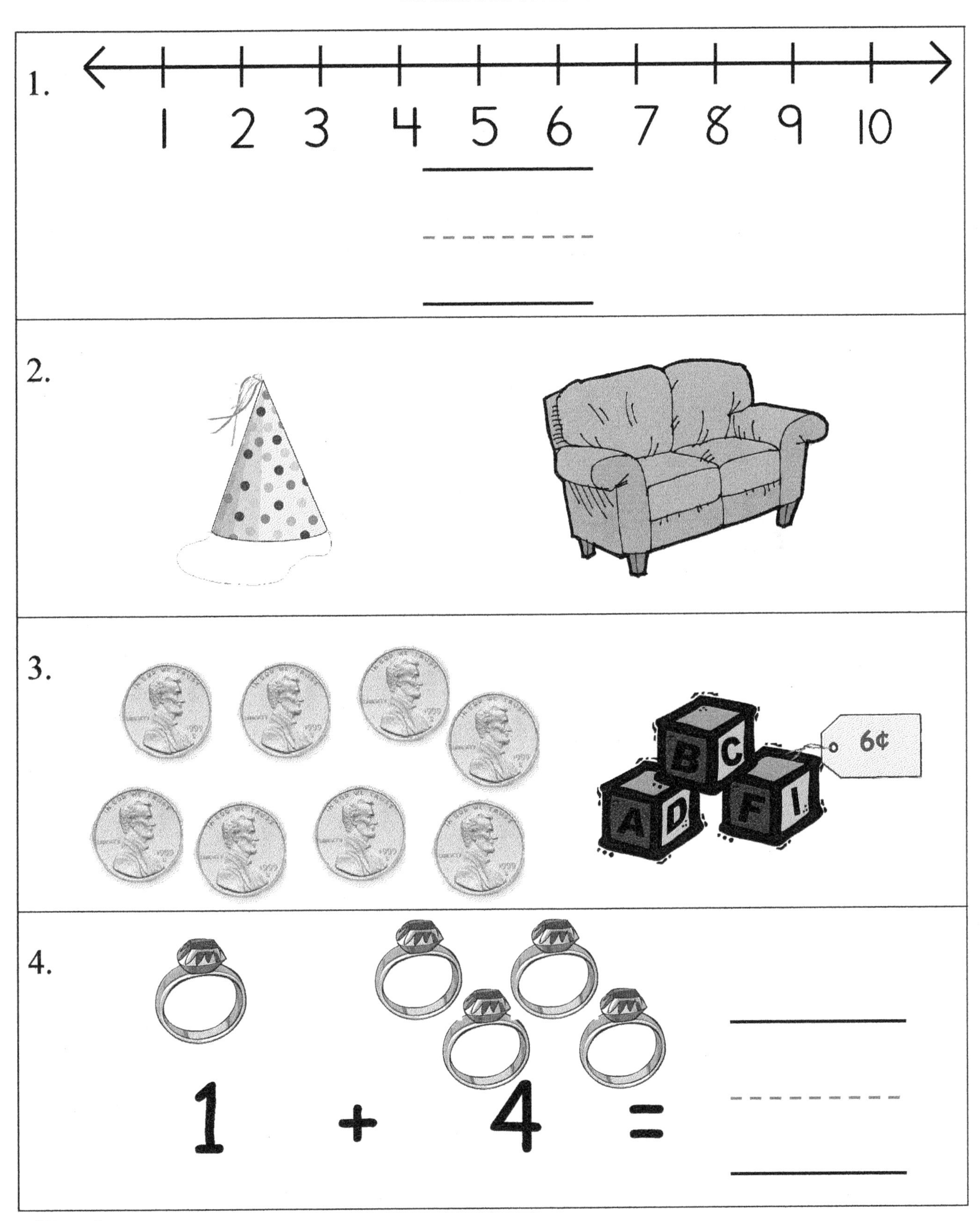

Directions:

1) Write the number after 7.
2) Circle the heavier one.
3) Circle the pennies used to buy the toy.
4) Write how many rings in all.

Lesson #137

1.

4 ____ 6

2.

2 + 1 = ____

3.

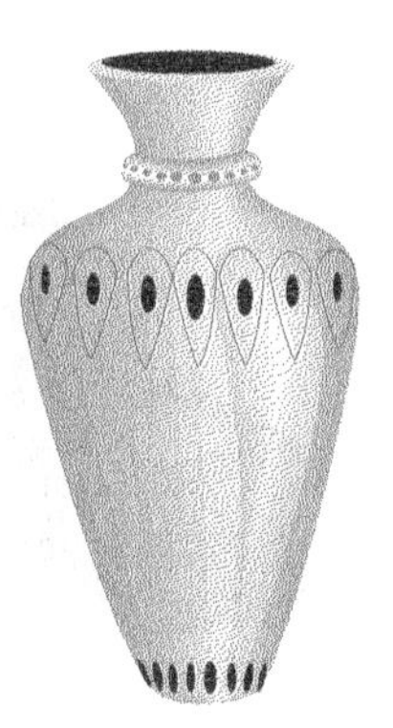

4.

____ ¢

Directions:

1) Write the missing number.
2) Write how many bears in all.
3) Circle the one that holds less.
4) Write how many cents.

Lesson #138

Directions:

1) Circle the pennies used to buy the candy.
2) Count the number of mice. Write the number.
3) Write the time shown on the clock.
4) Write how many deer in all.

Lesson #139

1.

3 4 ___ 6 ___ 8

2.

JULY

Sun	Mon	Tue	Wed	Thu	Fri	Sat
					1	2
3	4	5	6	7	8	9
10	11	12	13	14	15	16
17	18	19	20	21	22	23
24	25	26	27	28	29	30
31						

3.

1 + 3 = ___

4.

8¢

Directions:

1) Write the missing numbers.
2) Circle every Wednesday in red.
3) Write how many buses in all.
4) Circle the pennies used to buy the scissors.

Lesson #140

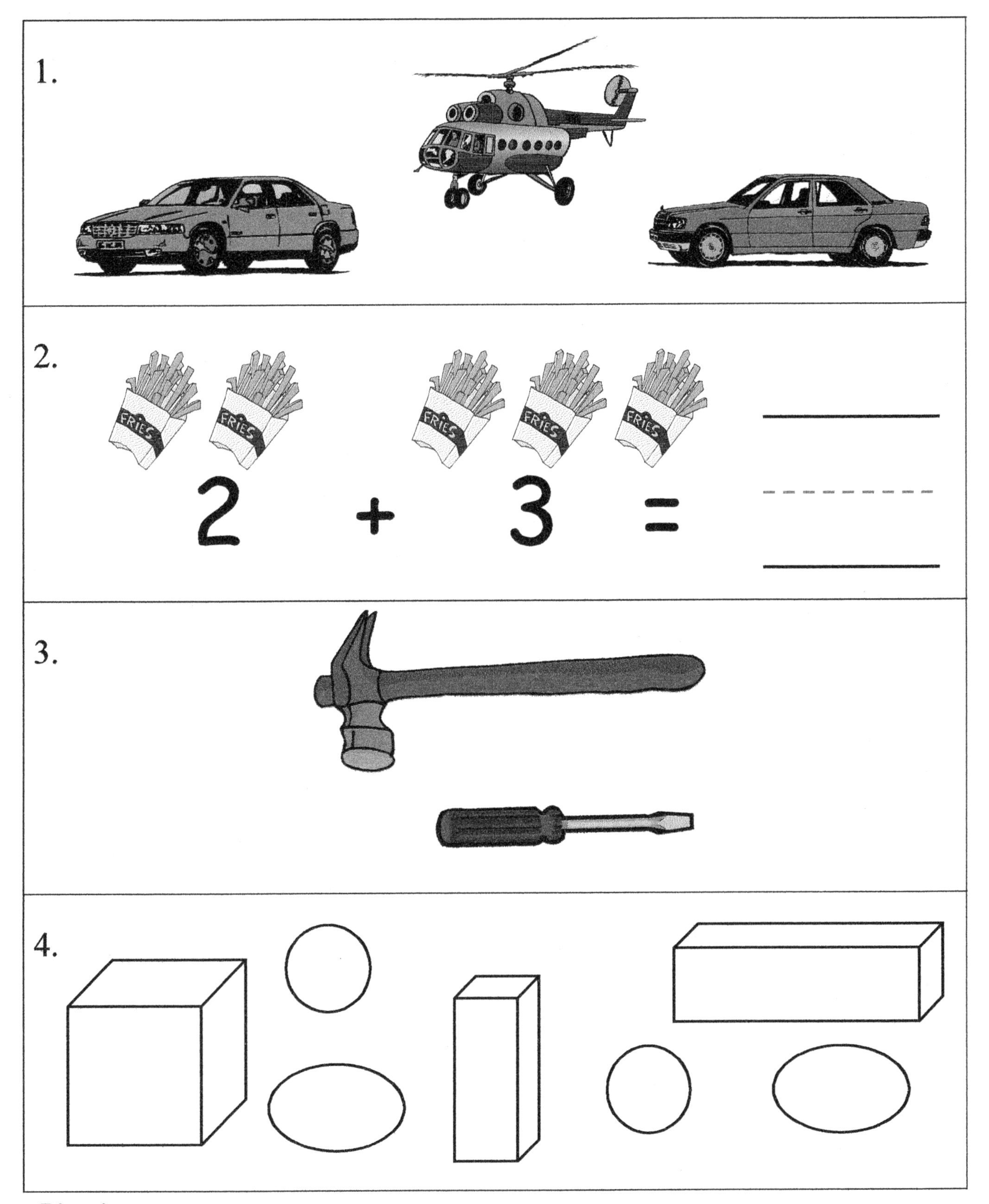

Directions:

1) Circle the two that weigh about the same.
2) Write how many orders of fries in all.
3) Circle the shorter one.
4) Color the shapes with corners in red and the shapes without corners in green.

Level K

2nd Edition

Mathematics

Help Pages

Help Pages

Numbers	
0	zero
1	one 
2	two 
3	three 
4	four 
5	five 
6	six 
7	seven 
8	eight 
9	nine 
10	ten 

Help Pages

Comparing

Before/After:

The number that is **before** 3 is **2**.

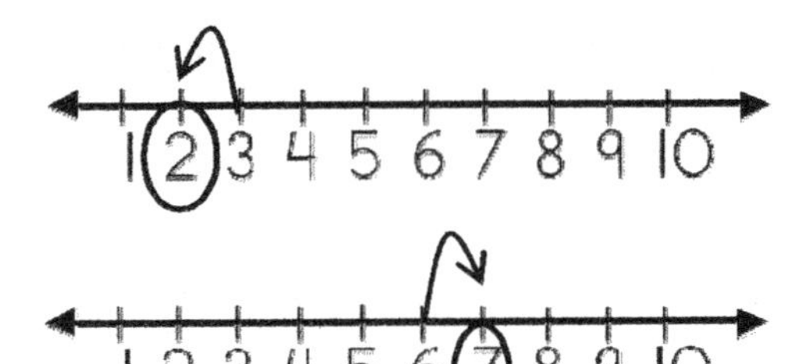

The number that is **after** 6 is **7**.

More/Fewer:

There are **5** dogs in this group.

There are **3** monkeys in this group.

This group has **more**.

This group has **fewer (less)**.

The number **5 is more than 3**.

Longer/Shorter:

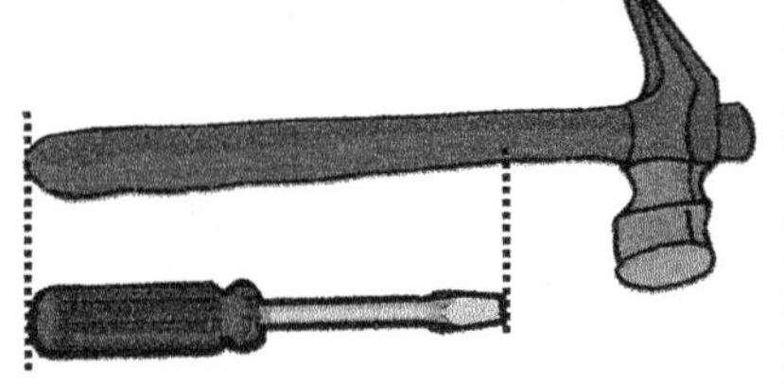

The hammer is **longer** than the screwdriver.
The screwdriver is **shorter** than the hammer.

Taller/Shorter:

The lamp is **taller** than the candle.
The candle is **shorter** than the lamp.

Help Pages

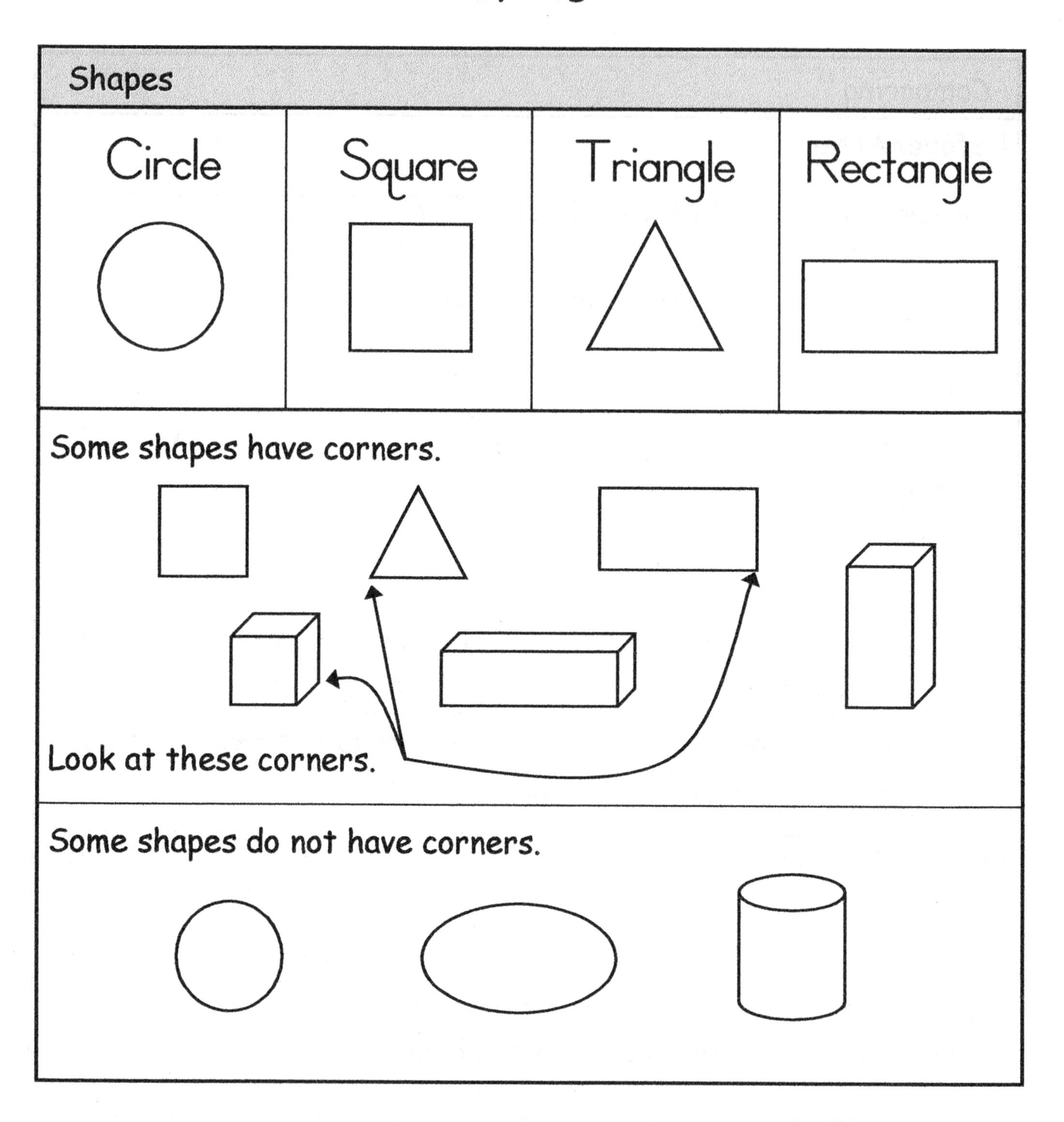
Shapes
Circle
Square
Triangle
Rectangle
Some shapes have corners.
Look at these corners.
Some shapes do not have corners.

Help Pages

Money

This is a **penny**. It is worth **1¢**.

This is a **nickel**. It is worth **5¢**.

This is a **dime**. It is worth **10¢**.

Time

A clock has two hands. The long hand is the **minute hand**. The short hand is the **hour hand**.

The minute hand is pointing to the 12.

On this clock the hour hand is pointing to the 4.

The time is **four o'clock** or 4:00.

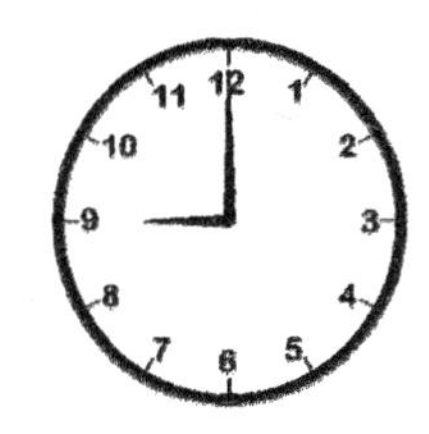

The time is **nine o'clock** or 9:00.

Help Pages

Calendar:

Reading a Calendar:

June

Sunday	Monday	Tuesday	Wednesday	Thursday	Friday	Saturday
		1	2	3	4	5
6	7	8	9	10	11	12
13	14	15	16	17	18	19
20	21	22	23	24	25	26
27	28	29	30			

- A calendar tells the **name of the month**. This month is **June**.
- The **days of the week** are at the top of the calendar. There are **7** days in a week. The days of the week are: **Sunday, Monday, Tuesday, Wednesday, Thursday, Friday, and Saturday**.
- A calendar also tells how many days are in the month. This month has **30** days.

Help Pages

Addition:

When you find **how many in all**, you **add**. The plus sign (+) tells you to add.

2 tigers + 3 tigers = 5 tigers

There are **5 tigers** in all.

1 duck + 2 ducks = 3 ducks

There are **3 ducks** in all.